Zohrab Makiyan

Female Sexuality

Essay

Zohrab Makiyan

Female Sexuality

Essay

Contents

Omnia vincit amor, et nos cedamus amori (latin)
Love wins all, and we submit to love

Publius Vergilius Maro, 70-19 BC

Introduction

Sexuality is the range of psychic reactions and emotions connected with sexual desire manifestations/satisfaction and, ultimately, the reproductive instinct, or the instinct of *procreation*.

Females usually have the ability to realize reproductive functions very early; but at this stage, not always ready for realization of sexuality.

Anorgasmia and *infertility* are not same, they are different problems.

Female sexual disorders (i.e., sexual dysfunctions, disharmony, vaginism, dyspareunia or anorgasmia) may be a direct or indirect cause of divorce in nearly 45% of married couples [2, 5, 8, 13]. Impaired satisfaction is one of the reasons for frequent changing sex partners in pursuit of sexual pleasure and experience. Quite often, however, individual problems of erogenous sensitivity or satisfaction remain unresolved. Impaired satisfaction often accompanied by depression, sleep disorders, decreased libido or neurosis [2, 8, 11, 13, 19].

Sexual dysfunctions are mainly treated by urology, sexopathology and psychiatry. Objective examination of disorders of sexual function should be based on gynecological examination, not only discussing psychological problems.

Deepening knowledge in female sexology is an important element of the development of gynecology.

This monograph presents up-to-date data on the anatomy of female genital organs and female sexual functions, deviations and disorders, as well as methods of sex therapy; also, some aspects of psychoanalysis are discussed.

Potentially, the experience gained from this book will be helpful in reaching sex life harmony.

Zohrab Makiyan, 2013

1. Sexology

Sex (Latin: sexus) is the gender relations that are based on the instinct of procreation of the human species. The Latin word originally meant "division" and is related to the words "section" and "secator" – to cat.

Latin word "sex" is index number of **6**, thus, can be interpreted as meaning "the sixth sense" in addition to the five main senses, including vision, hearing, smell, touch and taste.

In biology, the term "sex" is used to distinguish between two types of individual organisms and their organs of reproduction, including *sex cells*. In case of oogamy (i.e., sexual reproduction ensured by oocyte and sperm cell), the terms "female sex" and "male sex" are used respectively. Also, they are identified using the signs "+" and "-".

Sexology (sex + the Greek word "logos" meaning "science") is the scientific study of physiological, psychic and social aspects of relations between the sexes. Sexological studies cover all manifestations of human sexuality and sexual functions, including normal sexuality and various sexual practices, among them the so-called "paraphilias" or "sexual deviations". In a broad sense, sexology is a multidisciplinary field of knowledge, based on the integration of physiological and pathologic data's from various medical disciplines, including psychiatry, endocrinology, gynecology, andrology, neurology, etc.

Sexopathology (*sex* + the Greek word *"pathos"* meaning *"feeling"*, *"emotion"*, *"suffering"* or *"disease"* + the Greek word *"logos"* meaning *"thought"* or *"science"*) is the field of clinical medicine that covers sexual disorders and develops methods for their diagnosis, treatment and prophylaxis. Sexopathology as a part of medical sexology, studies functional and emotional aspects; personal and social related factors, of the sexual disorders that constitute the main group of sexual pathologies.

Sexuality is the feature, characteristic of sexual desire, sexual reactions, sexual activity and sexual fantasies in humans; it includes the whole range of motivations, sets and behaviors aimed at sexual instinct realization.

Sexual function (*sexuality*) includes male genetic material (i.e., spermatozoa) introduction into female genital tracts for the purpose of child-birth, or **procreation**.

Procreation (the Latin word "procreatio" meaning "birth"; synonyms: "reproduction", "reproductive function") is the production of *progeny* (the Latin word "pro" meaning "forward" and "creatio" meaning "creation"); this is the function of reproduction, or multiplication.

Sexual response consists of the following four phases: arousal, plateau, orgasm and satisfaction; it is produced of coordinated functioning of somatic and autonomous nervous systems innervating genital organs.

Disorders of sexual function, or *sexual dysfunctions,* can manifest themselves by decreased libido, sleep disorder, anxiety, depression and/or dysphoria. Often sexual dysfunctions are caused by *psychogenic factors.*

Penetration, Copulation (Latin: *coitus*) - genital contact between male and female partners for the purpose of reproduction and (in humans) for reaching satisfaction/sexual pleasure. Synonyms include *sexual intercourse, copulation, sexual act, sexual relation, penetration.*

Figure 1. A fragment of famous Khajuraho temple in India, erotic stone carving (9th-12th century AD).

2. History of sexology

Historical review of sexology makes it possible to trace the evolution of thinking dispositions, knowledge and preferences in the intimate domain.

Sexology separated of other scientific disciplines to become a substantive field of scientific study only in the 20th century, with growing demand in knowledge of sexuality manifestations. Problems of sex and intimate relationship have always attracted interest among investigators.

Ancient myths and, at a later stage, philosophical teachings contained definite information about the nature of sex differences, anatomy and physiology of genital organs, coital techniques, conception, pregnancy and delivery. However, ancient *erotology,* or theory and practice of the art of love, did not have sexuality studies as its aim.

Eros (Ancient Greek: *Eros,* or *Amour,* Roman: *Cupidon*) is God of love in the Ancient Greek mythology; he is Aphrodite's permanent companion and assistant, the personification of sexual desire ensuring the continuation of life on Earth. *Eros* a global God, uniting other gods as marital couples, was born of "Chaos" - Dark Night and Light Day, or Sky and Earth. He governs both the environment and the internal morals of folks and gods by controlling their hearts and passions. In relation to nature, he is a benevolent god of spring, who fertilizes the earth and produces new life. He was thought to look like a beautiful boy with wings successively depicted as having a flower and a lyre, and then arrows of love or a flaming torch.

The Ancient Greek language contains the words showing the already existing differentiation between types of love:

- agape (Ancient Greek: ἀγάπη) meaning selfless/unconditional love that was later, in the Christian world, described as God's love for human beings;
- storge (Ancient Greek: στοργή) meaning tender affection, especially the familial one;
- philia (Ancient Greek: φιλία) meaning love of friendship (or love of attachment) caused by social ties and personal choice;
- eros (Ancient Greek: ἔρως) meaning spontaneous exalted amorous state in the form of admiration directed to object of love.

Pornography (Ancient Greek: *"fornicatress"* + *"to wright"*; *colloquial speech: "porno"*) means any naturalistic image, picture or description of genital organs and coitus aimed at producing sexual arousal.

The term *pornography* originated from the book "Le pornographe, ou idées d'un honnête-homme sur un projet de réglement pour les prostituées, propre à prévenir les

Malheurs qu'occasionne le Publicisme des Femmes" by *Restif de la Bretonne* that was first published in France in the XVIIIth century (1781). This book reviewed the domains of human life traditionally considered indecent by public standards, in view of which its title became another name for sexuality-related obscenity. Pornography, as a vulgar presentation of sexual topics, is often distinguished from erotics identified as an aesthetically satisfying presentation of the same topics. However, no generally recognized division between the two notions exists. Such division depends on culture, local laws, traditions and religion. According to the ordinary viewpoint, pornography, as opposed to erotics, is concentrated on presenting physiological aspects of coitus and human body-accented sexuality. Often, featuring an erected penis and open vagina is considered as a distinguishing trait of pornography.

Erotics, in contrast, is mostly concentrated on sensual aspects of interpersonal relations, and uses coitus as a means to present human body aesthetics in combination with emotions of the acting persons. Several ancient guides to the art of love can be found in the literature, including Ovid's "Ars Amatoria" (The Art of Love), the Ancient Indian treatises "Kama Sutra" by Mallanaga Vatsyayana [1] and Ananga Ranga, as well as the Ancient Arabic literary work "The perfumed garden of sensual delight".

The Ancient Indian text "Kama Sutra" (Sanskrit: कामसतर) written in Sanskrit is dedicated to "kama", the

11

domain of sensual and emotional life, desire and love [1]; its full title is "The Kama Sutra of Vatsyayana". Kama Sutra is the earliest Ancient Indian guide to manifesting the feeling of love.. The text was authored by Mallanaga Vatsyayana, a philosopher and physician; supposedly he lived in the 3rd-4th century AD. Mallanaga Vatsyayana traditionally regarded as a person who took an ascetic vow. In mass culture, this treatise is mostly known for its section (chapters) dedicated to sex positions; however, it mainly presents philosophy and psychology of relations between sexes, as well as advice for making advances toward girls, living a happily married life and extramarital relationships [1].

Kama Sutra has not lost its significance as a guide to sexuality nowadays. It was not an illustrated text, and presentation of the majority of sex positions was limited to short description or just naming; however, the Khajuraho temple erotic stone carving (Fig. 1) is thought to stem from this text.

In India, the rulers of Chandella Dynasty built 85 temples, of which only 22 temples have been preserved to the present day. Between 950-1050 AD, according to legend, in Khajuraho the god of Chandella built a temple with human passions carved in stone that brought visitors to understanding of naturalness of sexual desire. Khajuraho temples' sculptures are the world's most famous, most spicy and most open ancient monuments of this kind and the unique milestones of the history of sexology (fig. 1).

Xenophanes, an Ancient Greek philosopher (c. 570-478 B.C.), in his scriptures, considered human soul, similar to air, to have heavenly origin, and human body to belong to earth [176].

Plato expanded this notion to cover sexual life and developed a teaching on heavenly and earthly Eros to end up with a real erotic mysticism. He was the first to develop a teaching on sexual energy transformation, or *sublimation*, into psychic phenomena that was later developed further by *Nietzsche* and *Freud*.

According to *Plato*, each genuinely creative activity is merged with sensuality. He applied the same notion "creative activity in the domain of perfection" to both sexual activity and mental state and considered both of them as stemming from the very depths of the human personality. However, *Plato* valued the purely physical manifestations of sexual instinct less than its psychical element. Probably, this was due to the fact that the Ancient Greeks felt disgust towards any victory of sensual over rational and towards any enslaving of wit, such as that especially evident during coitus.

Democritus (ca. 460 – ca. 380 BC) kept a reserved attitude toward sexual pleasures, considering them as "similar to transient apoplectic paroxysm", but "able to cause the sensations comparable to those evoked by scratching the itchy places of the body".

Hyppocrates (ca. 460 BC) called sexual intercourse a "petit mal epilepsy" (falling sickness) to emphasize the presence of characteristic "convulsions and transient unconsciousness" associated with paroxysm. According to historians, Hyppocrates believed that human sperm has cerebral origin and reaches the genital ducts via the spinal cord.

Publius Ovidius Naso (born on March 20, 43 BC) was an Ancient Roman poet, famous for his works in many genres and, especially, for his love elegies and the two poems, "Metamorphoses" and "The Art of Love" (Latin origin - Ars Amatoria) [178].

Leonardo da Vinci (1452-1519) in 1493, using a pen and ink technique, produced the drawing "The Copulation" showing the anatomy of male and female genital organs during coitus. This drawing demonstrated the nervous and vascular bundles supplying genitals. The inscription placed by the author above the drawing copulation of a hemisected Man and Woman says: "*I expose men to the origin of their first, and perhaps second, reason for existing.*" - Leonardo Da Vinci

Systematic studies of sexual life were started by psychiatrists and gynecologists and initially covered pathologic forms of sexual activity rather than its normal forms. Forefathers of sexology include *Richard von Krafft-Ebing* (1840-1902), professor of psychiatry at the University of Vienna; *August Forel* (1848-1931), a Swiss neurologist, psychiatrist and entomologist; *Albert Moll* (1862-1939) and *Magnus Hirschfeld*

(1868-1935), German psychiatrists; *Sigmund Freud* (1856-1939), the father of psychoanalysis; Iwan Bloch (1872-1922), a German dermato-venerologist; *Henry Havelock Ellis* (1859-1939), an English publicist, editor and physician; and *Theodoor Hendrik van der Velde* (1873-1937), a Dutch gynecologist.

Considerable contribution to development of sexology was made by *Alfred Kinsey* (1894-1956), an American biologist (1894-1956); *Hans Gise* (1920-1970), a German physician; and *William Masters* (1915-2001), an American gynecologist.

Recent studies in genetics, endocrinology, neurophysiology, embryology, evolutionary biology, gynecology and other disciplines significantly enriched and expanded the knowledges in the field of differentiation and interrelations of sexes, as well as human sexuality manifestations.

Development of sexology was significantly influenced by literary masterpieces of scandalously famous writers whose books were enthusiastically read by representatives of European aristocracy in conditions of confidentiality.

Donatien Alphonse François de Sade (1740-1814), best known as *Marquis de Sade*, a French aristocrat, writer and philosopher, propagated absolute freedom not restricted by morality, religion or law and considered the satisfaction of personal desires and strivings to be the most valuable aspect of life. In the works of the German sexologist *Richard von Krafft-Ebing*, sexual satisfaction reached through causing pain and/or humiliation to other person was called after his name, sadism; later, the words "sadism" and "sadist" started to be used in a broader sense. The most famous literary works by Marquis de Sade include "The 120 days of Sodom, or the School of Libertinism" ("Les 120 journées de Sodome, ou l'École du libertinage"), a novel first published in 1785; first edition of "Unfortunes of virtue" ("Les infortunes de la vertu"), a novel published in 1787; and second edition of "Justine, or the unfortunes of virtue" ("Justine ou les malheurs de la vertu"), a novel published in 1788.

Historical facts about the life of Marquis de Sade include the following: "On January 5, 1772, Marquis de Sade invited some familiar noblemen to see the first performance of his comedy to be staged by himself in Lacoste, his family estate. At 10 o'clock in the morning, Marquis de Sade, together with his man-servant, climbed upstairs to the room of a girl named Borelli and nicknamed Mariette. Three other girls – Rosa Coste, Marionette Loget and Marianne Laverne – were also present in the room. According to the police charge sheet, these persons involved themselves, together with Marquis de Sade, into the following activities in this room: active and passive flagellation, anal sex (denied, however, by the girls) and eating excitatory sweets offered by Marquis de Sade (actually, these sweets contained Cantharis, an aphrodisiac causing harmful effects on human health).

Leopold Ritter von Sacher-Masoch (born on January 27, 1836 in Lvov (first called Lemberg) and died on March 9,

1895 in Frankfurt-on-Maine or Lindheim), an Austrian writer, addressed the theme of the dominant woman taking control of a submissive man in many of his works, even the historical ones. His descriptions were so impressive that, in 1886, a Vienna psychologist Richard von Krafft-Ebing proposed that the term "masochism" be used to denote the sexual pathology characterized by getting sexual pleasure from being abused or dominated.

Leopold was brought up in the home of his parents in a liberal enlightened atmosphere characteristic of the period of Franz Josef's reign. Already as a child, he showed certain inclinations that made him famous at a later stage of his life. Sacher-Masoch was attracted by episodes of cruelty; he liked to view pictures showing executions, and his favorite reading was the Book of Martyrs. The person who played an important role in development of his personality during childhood was the Countess *Zenobia*, his father's relative; she was an extremely beautiful and, simultaneously, cruel woman. Once, when playing hide-and-seek with his sisters, he hid himself in the Countess's bedroom and witnessed the Countess coming in the bedroom with her lover and caught in the act by her husband and two of his friends several minutes later. The Countess gave a good beating to the three unwelcome guests and expelled them; the lover escaped; and Leopold imprudently disclosed his presence, which resulted in him being beated also. This beating, however, caused the boy to feel a weird pleasure. Soon the Count returned, and Leopold hiding behind the door heard the crack

17

of a whip and the Count's moans. Humiliation, the whip and the fur that the Countess liked wearing became the constant attributes of Sacher-Masoch's works; since then, he regarded women as the creatures that should be simultaneously loved and hated. Sacher-Masoch's works were translated into many European languages and published in large editions. He enjoyed special popularity in France. His works were highly appreciated by Émile Zola, Gustave Flaubert, Alphonse Daudet, Alexandre Dumas, père and Alexandre Dumas, fils. In 1886, the first French President François-Paul-Jules Grévy personally decorated Sacher-Masoch with a Legion of Honour award.

One of the earliest investigators of sexuality who worked before the XXth century was *Richard Freicherr von Krafft-Ebing* whose book "Psychopathia Sexualis" published in 1866 contained the description of a very impressive range of sexual anomalies.

The first investigator proposing the concept of sexology as an individual scientific discipline was dermato-venerologist Iwan Bloch who emphasized in his work "The sexual life of our time and its relations to modern civilization" (1909) that human sexuality studies should be an integrated scientific discipline combining the evidences obtained in biology, medicine, anthropology, philosophy, psychology and ethnology, as well as the history of literature and art. Also, I. Bloch is the author of the well-known book "History of prostitution" [4].

At the end of the XIXth century and beginning of the XXth century, *Sigmund Freud* developed a sexuality theory based on his studies on patients. Sigmund Freud (full name – Sigizmund Shlomo Freud) was born on May 6, 1856 in Freiberg (Austrian Empire), now called Příbor (Czech Republic), and died on September 23, 1939 in London; he was an Austrian psychologist, psychiatrist and neurologist, the founder of a therapy-oriented psychoanalytical school in psychology. He postulated the theory of close interactions between the conscious and subconscious processes as the basis for development of neurotic disorders in humans [21, 22].

Theories and treatment methods used by Freud caused polemics in Vienna in the XXth century and remain hotly disputed subjects up to date. Apart from continuing discussions in scientific and medical literature, Freud's ideas are frequently discussed and analyzed in literary and philosophic works. He is frequently and quite reasonably called "the father of psychoanalysis". In 1930, S. Freud was awarded the Goethe Premium for his significant contribution to science; this award was a welcome support to him and stimulated further dissemination of psychoanalytical knowledge in Germany. In 1933, however, with the advent of Nazism (A. Hitler appointment as Chancellor), the situation sharply changed to unfavorable. The course of events was quite quick: a number of discriminatory anti-semitic laws were adopted, and the books contradicting Nazi ideology were burned or destroyed. The works of Freud, as well as the works of *Heine, Marx, Mann, Kafka and Einstein*, were prohibited. Princess Marie Bonaparte, S. Freud's favorite adherent and patient, managed to agree his release from jail with the Nazis after selling her private residence and paying a ransom of 100,000 shillings. Freud's family moved to London, but his four sisters were killed in gas chambers.

The scientist worked very intensively; the most famous of his works include "Beyond the Pleasure Principle" (1920), "Three essays on the theory of sexuality" (1920), "Mass psychology" (1921) and "I and it" (1923). In April 1923, S. Freud was diagnosed with palatal tumor. The first surgical operation was not successful, and he was close to death. This surgery was followed by a total of 32 similar surgical operations. In the summer of 1939, S. Freud's suffering from the progressive disease became especially severe. He reminded his treating doctor *Max Schur* about his promise to help him die. Initially, his beloved daughter Anna who spent all her time at his bed opposed this idea of her father, but then agreed to it seeing his suffering. On September 23, doctor Max Schur injected a lethal dose of morphine to S. Freud. Sigmund Freud died at three o'clock in the morning. His body was cremated in Golders Green, and his remains were placed in the ancient Etruscan vase that he had been presented with by Marie Bonaparte.

Marie Bonaparte (born on July 2, 1882 in Saint-Claud and died on September 21, 1962 in Saint-Tropez, France) was the grand niece of Napoleone Bonaparte and, according to her, "the last representative of the Bonaparte family". She was the great-granddaughter of Lucien Bonaparte, the brother of Emperor Napoleone Bonaparte. Marie inherited a large fortune from her grandfather on mother's side who was a successful businessman, one of the Monte Carlo developers. After marrying Prince George of Greece and Denmark (the second son of George I and Olga Konstantinovna) in 1907, she acquired the title Princess of Greece and Denmark.

Marie Bonaparte in 1925 became acquainted with Sigmund Freud who started her didactic psychoanalysis on September 30. Since 1925, she spent several months in Vienna where Sigmund Freud carried out her psychoanalysis sessions. In 1929, Marie Bonaparte finished her Sigmund Freud's psychoanalysis course and became one of his favorite students. Though the traditional duration of the analysis was several months, Marie Bonaparte remained S. Freud's *analysante* until 1938, when he had to quit Austria. She initiated the tradition of *"interrupted psychoanalysis"*, according to which the analysante living in another country regularly visits to own *psychoanalyst*, each visit lasting for several weeks. At present, this type of analysis is actively practiced by many psychoanalytical schools in France. On November 4, 1926, Marie Bonaparte established the first psychoanalytical society, the Psychoanalytical Society of Paris (Société Psychanalytique de Paris); at present, it is the most influential of the existing psychoanalytical societies. The most well-known work by Marie Bonaparte is "Female sexuality" ("De la sexualite` de la femme") [35], published in 1951.

Magnus Hirschfeld in 1908 had started to issue The Journal of Sexology, the first scientific journal dedicated to problems of sexology, and in 1918 he founded the first Institute of Sexology (*Institut für Sexualwissenschaft*) in Berlin. M Hirschfeld studied the medical, ethic and legal problems connected with relations between the sexes, including the questions related to homosexuality and birth control. M Hirschfeld conducted the First International Congress for Sexual Reform in Berlin in 1921 and founded the World League for Sexual Reform in 1928. One of the first actions taken by Nazis after their coming to power was the destruction of the Institute on May 6, 1933 and burning its library.

After 1950s, the center of sexologic studies activity displaced to the USA.

Alfred Charles Kinsey (born on June 23, 1894 and died on August 25, 1956) was an American biologist, professor of entomology and zoology, who founded the institute for research problem of sex, gender and reproduction. In 1947, Alfred Kinsey founded his *Institute for Sex Research*, now bearing his name, in the Indiana University (Bloomington). Zoologist Alfred Kinsey was one of the first scientists who placed sexuality studies on a positive basis. Kinsey's studies in the field of human sexuality caused a profound effect on the social and cultural values in the United States and many

22

other countries in the 1960s, paralleling the advent of the sexual revolution. His important achievements included, but were not limited to, the description of a range of sexual practices on the basis of numerous biographic interviews and questionnaire data and conceptualization of knowledge on the continuous character of sexual orientation (*Kinsey's Scale*). Kinsey showed that significant part of general population constantly deviate from the so-called normal sexual behaviors determined by social morals, and the majority of people deviate from it from time to time. Results of his studies were reported in two volumes of *Kinsey's Reports* (1948 and 1953) [23, 24].

At the end of the 1960s, further progress of sexology was manifested by publication of the works "Human Sexual Response" (1966) and "Human Sexual Inadequacy" (1970) by *William Masters* and *Virginia Johnson*; these books became bestsellers. In 1978, they founded the *Masters & Johnson Institute*. Masters and Johnson pioneered the experimental studies of sexuality on volunteers.

Fritz Klein developed a sexual orientation grid - a multi-dimensional system for detailed description of complex sexual orientation similar to one-dimensional Kinsey Scale, but individually measuring each of the seven different vectors of sexual orientation and identity and making it possible to describe temporal course of changes. In 1978, F Klein published "The Bisexual Option", an innovative psychological study of bisexuality, and, in 1998, he founded the *American*

Institute of Bisexuality (AIB) to encourage and support the bisexuality studies, as well as to provide the corresponding enlightenment in the issues of ethics and law.

Russian sexological studies had been carried out since the beginning of the XXth century and were interrupted at the beginning of the 1930s. Revival of the Soviet sexology in the 1970s was associated with publication of the first guides for physicians "General sexopathology" (1977) and "Specific sexopathology" (1983) authored by *G.S. Vasilchenko* [5], the organizer of the All-Union Scientific Methodological Center for Sexopathology currently named "The Russian Federal Center for Sexology and Sexopathology".

Some pioneer proceedings of Russian doctors dedicated to innovative sexology and family relations:

"Sexual abstinence in front of Tribunal of Medicine" (1906) and "Female sexual anesthesia" (1927) by gynecologist-sexopathologist *L.Ya. Yakobzon*;

"Clinical lectures" (issued in 1913) by *S.P. Botkin*. Sergey Petrovich Botkin (1832—1889) was a famous Russian clinician, therapist, and activist, one of the founders of modern Russian medical science and education. In 1874, he organized a school of nurse, and in 1876 began lectures "Women's medical courses".

S.P. Botkin had experienced in various parts of medicine: in the clinic of Professor Hirsch in Königsberg, in the pathological institute of R. Virchow in Würzburg and Berlin, in the clinic of the famous physician L. Traube, neurologist Romberg, doctor syphilidologist (specialist of syphilis treatment) Berenshprung in Berlin, in the laboratory experimental physiologist Claude Bernard in England.

His lectures about female sexual life include family hygiene, *modus vivendi* (latin, means – manner of life) and normal sexual rhythm.

He was the court physician for both Tsar Alexander II and Tsar Alexander III. He was the father of Dr. Eugene Botkin, the court physician for Tsar Nicholas II (before Soviet Revolution, 1917).

"Sexual practicies" (1926) by *V.M. Bekhterev*. "Foundations for Brain Functions Theory" (1903), described Bekhterev's views on the functions of the parts of the brain and the nervous system. It also suggested the Energetic Inhibition Theory which describes automatic responses - *reflexes*. This theory claims that there is an active energy in the brain which moves towards a center, and when this happens, the other parts of the brain are left in an inhibited state. Bekhterev's research on associated responses would become highly connected with the important area of psychology called *Behaviorism*. Objective psychology is based on the principle that all *behavior* can be explained by objectively studying reflexes. It also led to a long-standing rivalry with academician Ivan Pavlov.

Academician *Ivan Pavlov* (1849–1936) was a famous Russian physiologist. Pavlov contributed too many areas of physiology and neurological sciences. Most of his works involved researches in temperament, conditioning and unconscious reactions. *Carl Jung* continued Pavlov's work and correlated the introverted and extroverted temperament types in humans. Introverted persons, he believed, were more sensitive to stimuli and reached a *transmarginal inhibition* state earlier than their extroverted counterparts. This continuing research branch is gaining the name highly sensitive persons. *William Sargant* and others continued the behavioral research in mental conditioning.

As Pavlov's work became known in the West, particularly through the writings of John B. Watson, the idea of "conditioning" as an automatic form of learning became a key concept in the developing specialism of comparative psychology, and the general approach to psychology that underlay in behaviorism.

Pavlov's concepts about: reinforcement, unconditional and conditional reflexes, classical conditioning and instinct - was of huge influence to how humans perceive themselves, their behavior and learning processes and his studies of classical conditioning continue to be central to modern behavior therapy. The British philosopher Bertrand Russell was an enthusiastic advocate of the importance of Pavlov's work for philosophy of mind [Russell, Bertrand (1931), *The Scientific Outlook*, London: George Allen & Unwin]. Pavlov was awarded the Nobel laureate" in 1904, for recognition of his work in medicine and physiology.

During the Soviet period, well-known popularizing works were published by sociologist and sexologist *I.S. Kon* [20]; psychiatrist, psychotherapeutist and sexopathologist *A.M. Svyadoshch*; psychiatrist and sexopathologist *D.D.Yenikeeva*; and psychiatrist, sexopathologist and investigator of transsexuality *A.I. Belkin*.

Famous works of Russian sexologist about masculine impotence and female anorgasmia, frigiditats: "Male sexual disorders" (1960) by *I.M. Porudominskiy*; "Impotence" (1965) by andrologist *L.Ya. Milman*.

Large contribution to the development of modern sexology in this country was made by *Aleksandr Olimpievich Bukhanovskiy,* doctor of medicine, Professor of the Judicial Faculty of the Rostov State University and Head of the Department of Psychiatry and Narcology of the Rostov Medical University (born on February 22, 1944 in Groznyi, the USSR). Professor Bukhanovskiy is member of American Academy of Forensic Sciences, and member of American Academy of Psychiatry and Law and Honorary Member of the European Psychiatric Association. In 1969, he defended his PhD thesis dedicated to genetics of schizophrenia. He was the first Soviet psychiatrist to start the studies on the problems of sex change in transsexuals in 1980. Professor A.O. Bukhanovskiy published numerous topical scientific psychopathologic works on the basis of unique clinical material; he examined more than 500 transsexual patients. He is one of the founders of the new scientific discipline *Criminal Psychiatry* [2].

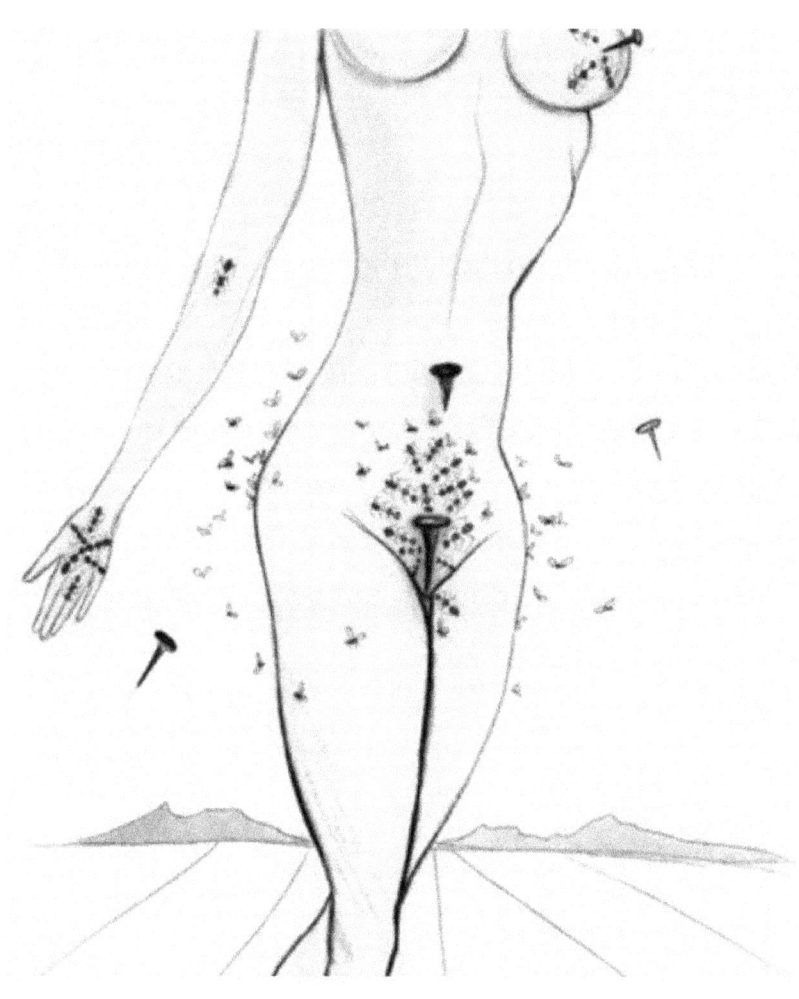

Ants-Nails-Flies-on-Nude

Salvador Dali. Engraving in Giacomo Casanova's novels

<a href="http://www.dali-genius.ru/">«Сальвадор Дали: XX век глазами гения»</a>

"They were both naked, the man and his wife, and were not ashamed.."

"Giving way to the temptation and in violation of God's will, she took of the fruit thereof, and did eat, and gave also unto her husband with her (The Book of Genesis 3:6). As a result, Adam and Eve learnt good and evil, realized their nudity and became ashamed (The Book of Genesis 3:1-7)."

The Book of Genesis

3. Anatomy and innervation of female sexual organs

Sexuality includes the reproductive function, that ensuring procreation - the main mission of every individual.

From the viewpoint of evolution and species survival, all functions of organism have one dominating purpose − to ensure the leaving of descendants (childbirth).

To ensure the fulfillment of this most important function, evolution has developed the mechanisms that support appropriate functions and, especially, dominate for realization of procreation (reproduction, childbirth).

How can motivate of individual for reproduction?

Obviously, the "pleasure principle" or the activation of mechanisms encouraging reproductive behavior is one of the corresponding instincts (for more details see Chapter 4 "Sexual function" and Chapter 9 "Psychoanalysis and the pleasure principle").

According to Russian academician I.P. Pavlov, survival of individual organisms and the whole species is ensured by a range of *involuntary reflexes* - instincts (or innate behavior) [10].

Interactions between central and peripheral systems results in the following:

- ✓ imperative desire for sexual contact (i.e., the so-called "sexual hunger"),
- ✓ intimate contact involving sexual arousal; readiness to initiate coitus (sexual intercourse) manifested at the psychosomatic level by vaginal moisturizing, clitoral or penile erection and increased blood circulation in the genital organs - in clitoris, vagina, uterus and ovaries; and vegetative reactions (i.e., increased heart rate and dilatation of peripheral vessels manifested by reddening of the face, neck, forearms and other regions),
- ✓ orgasm (sexual satisfaction) occurring during sexual intercourse at the peak level of arousal and accompanied by intense release of endorphins - "hormones of pleasure" and ecstatic euphoria,
- ✓ refractory phase following orgasm and lasting from several minutes or hours to several days.

The feeling of satisfaction associated with execution of reproductive function can be considered as a mechanism of "encouragement".

From the anatomical viewpoint, genital organs can be distinguished into:

- • the reproductive organs (ensuring childbirth)

- the erotic organs of sexual sensitivity.

Female reproductive organs:
  ➢ Ovaries: contain germ cells (i.e., oocytes) and to serve endocrine function,
  ➢ Uterus: it is a hollow organ, for gestate intrauterine development one or two (several) fetuses.
  ➢ Vagina: it is analogous to oviduct of mammals and ensures childbirth.

Organs of Erotic sensitivity:
  ➢ Clitoris
  ➢ Pudendal labia majora and labia minora
  ➢ Vestibulum Vaginae (Vulva)
  ➢ Erogenous zones (sensitive zones of the body).

Below is the brief anatomical information provided to ensure better understanding of specific features of innervation, erogenous sensitivity and sexual function.

Internal genital organs (Fig. 3.1)

Female genital organs are divided into internal and external genital organs.
Internal genital organs include: ovaries, Fallopian tubes, uterus and vagina.

External genital organs include labia majora and labia minora, clitoris and vaginal vestibule [3, 14, 16, 19].

Ovary *(ovarium)* is a paired organ that fulfills both reproductive function - production of female sex cells, oocytes; and endocrine function - secretion of sex hormones.

The ovary has oval shape and quite solid (tight) consistency. Its size and dimensions vary significantly depending on the age and the ovarian parenchyma functional status. On an average, its length, width and thickness are respectively 3-4 cm, 2-2.5 cm and 2-2.5 cm. Surface of the ovary is covered by a white coating, *tunica albuginea*, formed by dense fibrous connective tissue covered by mesothelium. It is colored light-gray and has glossy uneven surface.

The ovary consists of the *cortical* substance located under the ovarian coating and the *medullary* substance located deeper than the cortical substance. The cortical substance originates from *genital ridge* - germ cells. The medullary substance develops as a result of ingrowth of primary kidney tubules (mesonephros) into genital ridge substance.

Cortical substance contains numerous follicles of all degrees of maturity (primordial, preovulatory and ovulatory follicles) surrounded by connective tissue stroma. An ovary of a newborn female contains 300,000 to 400,000 oocytes. Beginning of follicle maturation coincides with the onset of puberty. During life of female, no more than 500 mature follicles are produced; the rest of the follicles are reduced.

The follicle maturation process involves proliferation of epithelial *Granulosa* cells, follicle growth and enlargement, and fluid-filled folliclular cavity formation. The mature follicle called *Graafian follicle* is about 2 sm in diameter. Approximately in the middle of the menstrual cycle, the mature follicle ruptures, and the follicular fluid containing the oocytes is released into the abdominal cavity; this process is called "*Ovulation*". The ruptured Graafian follicle transforms into the *corpus luteum*. In case of pregnancy, the *corpus luteum* functions as a temporary endocrine gland throughout the period of its persistence until 16-20 weeks gestation. In case of the absence of fertilization, the yellow body undergoes to atresia. Menstrual function is closely related to ovulation.

The ovaries produce sex hormones, including estrogens, progesteron and few amounts of androgens. Estrogens (estradiol, estriol and estrone) are present in the growing follicle cavity. Estrogens cause effects on the development of female genital organs, development of uterine mucosa - *endometrium*, and proliferation of vaginal epithelium. Progesterone is produced by cells of granulosa and *corpus luteum*; it causes endometrial secretory changes characteristic of the second phase of the menstrual cycle.

The menstrual cycle consists of the following three phases: *desquamation, proliferation and secretion.*

The desquamation phase is, on an average, 3 to 7 days long by endometrial withdrawal and menstrual discharge caused recession of low hormone levels.

This phase is followed by the proliferative phase, which lasts for about 14 days; during this phase, the dominant follicle grows to its maximum on the background of increased levels of follicle-stimulating hormone (FSH). The proliferative phase ends at the same time with ovulation (i.e., the release of the oocyte from the mature follicle). Follicular growth is accompanied by release of estradiol and endometrial proliferation. The ruptured follicle, resulting of ovulation is filled with blood and then with corpus luteum; in case of absence of conception, the corpus luteum persists for 2 weeks (14 days). Thus, the menstrual cycle lasts for 28 days (normal range – 21 to 45 days); 7 days of desquamative phase, 9-14 days of proliferative phase and 14 days of secretion phase. A new menstrual cycle then begins thus producing a new mature follicle.

Menstrual function normally begins at 12-16 and ends at 45-50 years of age. During menopause, the ovaries decrease in size; their hormonal function gradually disappears in the course of 5 to 7 years.

Vascularization. The ovary is perfusely vascularized; the blood vessels have spiral course and numerous branches. Distribution of the vessels depends on the phase of the menstrual cycle. A vascular plexus is formed in the developing internal coating (theca interna) during the period of growth of primary follicles, and degree of complexity of this plexus increases by the time of ovulation and formation of corpus luteum. Thereafter, gradual reduction of the vascular plexus is noted in parallel with regression of the corpus luteum. Veins are connected with each other via numerous anastomoses (connections), and venous vascular system capacity is significantly higher than the capacity of the arterial system.

Innervation. The parasympathetic and sympathetic nervous fibers entering the ovary form plexus around the follicles and corpus luteum, as well as in the medullary substance. The ovaries contain numerous receptors to send the afferent signals reaching the central nervous system and, in particular, to *hypothalamus.*

Uterine tubes (Fallopian tubes), or oviducts, are the paired organs connecting abdominal cavity with the uterus (Fig. 3.1). The oocyte passes from the ovary to the uterus via the uterine tube. The wall of the uterine tube consists of mucous, muscular and serous layer. Approximate length of the uterine tube ranges from 6 to 10-12 cm. Medial end of the Fallopian tube passes within the uterine wall and opens into the uterine cavity, and its lateral end widens to form an infundibulum with *fimbria* located adjacent to the ovary.

Usually, the fertilization process takes place in the uterine tube. The fertilized oocyte passes along the uterine tube due to contractions of the uterine tube muscular coat and fluctuating movements of cilia on the ciliated epithelium that lines the uterine tube cavity. Both endocrine and nervous factors can influence on the functional status of the uterine tube musculature. Also, various inflammatory processes in the uterine tube can prevent access of the sperm to the oocyte and deprive the woman of the ability to *conceive*.

Human uterus is a hollow pear-shaped muscular organ that houses the developing fetus.

The uterus is located in the pelvic cavity between the urinary bladder and rectum (Fig. 3.1). There is a wide variation in uterine dimensions between adult women. Uterine length ranges from 6 to 8 cm, uterine width at the level of fundus – from 3 to 5 cm, and transverse dimension – from 4 to 7 cm. The uterus consists of *fundus* (the wide upper part), *body* (the middle part) and *cervix* (the lower part). The narrow transitional zone between the body and cervix is called the uterine *isthmus*. The cervix protrudes into the vagina.

The wall of the uterus consists of mucous - *endometrial*, muscular – *myometrial*, and serous - *perymetrial* coats. The internal, endometrial coat is a mucous membrane lined by cylindrical epithelium. The mucous membrane contains the glands releasing their secretions into the uterine cavity. The uterine glands secretions are rich in mucin, have mucous character and pass into the vagina via the cervical canal. Cyclic endometrial changes (i.e., proliferation, secretion and desquamation phases) depend on sex hormone secretion and, accordingly, correspond to menstrual cycle phases.

Vascularization. Uterus is intensively vascularized with arteries and venous plexus, perfusely supplied with blood. The arteries supplying myometrium and endometrium pass spirally in the circular layer of myometrium, which contributes to their compression during uterine contractions.

Innervation. Uterus is richly innervated by <u>autonomous</u> nervous system (Fig. 3.2). A group of large ganglions containing both autonomic (vegetative) nervous cells and somatosensory cells are located in the parametrial adipose tissue near the uterine cervix.

Sympathetic nerve fibers into the uterus are supplied by the pelvic plexus connected with the celiac plexus. Parasympathetic innervation is ensured by the pelvic nerves. These nerves are formed with participation of the third sacral nerve and, sometimes, with participation of the second or the fourth sacral nerve. Branches of the pudendal nerves pass within the pelvic plexus. They innervate both external and internal genital organs (previously, it was thought that the pudendal nerves innervate only external genital organs). These somatic nerve fibers transmit the afferent impulses, including the impulses of pain. Stimulation of various sensory nervous structures causes shifts in the uterine functional status and influences on many functions of the body, including blood pressure regulation, heart rate, breathing, general metabolism, hormone production by hypophysis and hypothalamus activity.

Vagina (the Latin word *"vagina"* means "sheath"; the Greek word *"colpos"* means "depth") is a musculo-fibrous tube measuring 7 to 10 cm in length. Usually, its posterior wall is 1.5 to 2 cm longer than the anterior wall; vaginal width is 2 to 3 cm. The distal part of the vagina is separated by hymen from the vaginal vestibule. The vaginal vault contains the vaginal portion of the cervix that protrudes into the vaginal cavity between the anterior and posterior vaginal fornixes. The posterior fornix is deeper than the anterior fornix (Fig. 3.1).

Vaginal walls consist of mucous, muscular and external (adventitial) layers. The anterior vaginal wall is located immediately adjacent to bladder wall, and the posterior vaginal wall is adjacent to the rectal wall.

Vaginal mucous membrane is lined by stratified squamous epithelium. Its morphology changes depending on the period of a woman's life. Some amount of mucous discharge can be found in the vaginal cavity. This vaginal discharge is produced due to transudation from the mucous membrane capillaries and the addition of small amount of mucouses from the uterine cervix.

Vaginal mucous membrane has folds (Latin: *ruga, rugarum*). They are especially expressed in the lower part of the vagina and the posterior vaginal wall where they form the fold complexes, *columnae rugarum*. Due to its numerous folds, the vagina is characterized by significant stretching ability; this is especially important for fetal passage through the genital tract. The mucous membrane is more smooth in the parous women as compared to nulliparous women. The vaginal wall contains interwoven longitudinal and transverse muscular fibers. Degree of vaginal musculature development can be very high, which makes it able to produce intensive contractions. By using voluntary vaginal contraction, some women are able to squeeze the penis introduced into the vagina.

The vaginal wall is able to contract involuntarily in response to stimulation by oxytocin, a hormone stored in the posterior lobe of the pituitary gland, as well as a reflex response. There is a wide variation in resting-state vaginal depth between women. In the women that we have studied, it ranged from 6-8 cm to 11-12 cm. Simultaneously, the vaginal walls demonstrate significant stretching ability and elongate by 2 to 4 cm, if pressure is applied to the posterior fornix. Close correlation between the woman's height and the dimensions of her vagina has not been revealed. Vaginal discharges have mucous character and, if good personal hygiene is maintained, are not copious, sufficient just for moisturizing the vaginal mucosa.

To a certain degree, vaginal sexual function is influenced by pelvic floor muscles. They are able to concentrically narrow the vaginal opening, contribute to clitoral erection and squeeze the vestibular glands [147-151, 162].

Vascularization. Vagina is plenty supplied with blood from various sources, including: the vaginal branch (vaginal arteries) originating from the internal iliac artery; descending branch of the uterine arteries; and *aa. rectalis media* and *aa. vesicalis inferior* (*a. pudenda interna*) originating from the internal iliac artery. Venous blood is drained into *plexus venosus vaginalis* and, thereafter, into *v. iliaca interna*. Lymph vessels transport the lymph to *nodi lymphatici inguinales interni* (from the upper part of the vagina) and *nodi lymphatici inguinales* (from the lower part of the vagina).

Innervation (Fig. 3.2). Vagina is innervated by the vegetative nervous system – *plexus hypogastricus inferior* (sympathetic innervation) and *nn. splanchnici pelvici* (parasympathetic innervation). Vagina is supplied by branches originating from the common uterovaginal plexus (*plexus uterovaginalis*) whose lower anterior parts are called the vaginal plexus. Uterine and vaginal plexuses are composed of a relatively connecting nervous branches and ganglias of various sizes and shapes. The nerve branches surrounding the vagina reach the vaginal wall and form a plexus enclosing the ganglia of smaller size. The vaginal part of the joint uterovaginal plexus is connected with the plexus of the bladder and the rectum, as well as, via the uterine plexus, the plexuses of the ovaries and the nerves surrounding the uterine tubes.

The distal (i.e., vestibular) part of the vagina is innervated by the somatosensory *nervus pudendalis*.

External genital organs (Fig. 3.3)

Pudendal labia majora are the two folds of skin that form the borders of the pudendal swelling. They extend downward and backward from the mons pubis and are connected by the anterior and posterior comissures. External surface of the large labia majora is covered by hair, and their internal surface is adjacent to *pudendal labia minora.*

Growth of hair on the mons pubis and the pudendal labia majora reflects, to a certain extent, the degree of sexual maturation. During childhood, only the lanugo is present. Hair growth starts during the prepubertal period (11.5 to 12.5 years of age); mean length of the hair reaches 1 to 2 cm at the age of 13.5 years and 3 to 4 cm at the age of 14.5 years. In adult women, the hairs are slightly spiral, and the hair growth zone has triangular form, with the base of the triangle corresponding to the horizontal border of the hair growth zone, and the apex of the triangle directed to the perineum.

Pudendal labia minora are two thin folds of skin located immediately adjacent to the vaginal orifice (Figs. 3.1 and 3.3). Their external surface adjacent to the corresponding labia majora is covered by squamous epithelium similar to that covering the skin, but without hairs; and their internal surface is covered by mucous membrane and richly supplied with sensory nerve fibers.

The space bounded by the small pudendal lips is called *vaginal vestibule (*lat.: *vestibulum vaginae).*

Urethra and excretory ducts of the vestibular glands (also called *Bartholin's glands*) open into the vaginal vestibule. The fluid secreted by these glands moistens the vaginal opening.

Vascularisation. Pudendal labia minora are elastic and contain a tight venous plexus ensuring some erectility of these lips.

Innervation. labia minora, especially their internal surfaces, are richly supplied with nerve endings and quite sensitive to external stimuli. Free branching nerve endings and tactile nerve bodies are located in the epithelium, and incapsulated genital bodies are found in the dermis. Also, lamellated sensory bodies are located in the epithelium of the pudendal labia.

Vaginal vestibule, labia minora and clitoral head are jointly called V*ulva*.

Vulva is derivates from urogenital sinus: it develops from the distal part (hindgut) of embryonal primary intestine, called - cloaca. So, morphology and innervation of Vulva differs of labia majora and vagina.

Hymen is a pleat of mucosal membrane, located at the entrance to the vagina and separating the vagina from the vulva (Fig. 3.3). Hymen only partly close the entrance of the vagina: it has an opening admitting tip of the little finger. The structure similar to the human hymen is present in primates and some carnivores. The physiological role played by the hymen is unclear. Possibly, this membrane prevents infections invasion into the vagina. Hymens vary in shape. Most often, the hymen is ring-shaped, with a round or oval opening; less often, it can be semilunar, fringed, lobular, flappy, serrated, etc.

Hymen integrity is usually violated during the first sexual intercourse. Hymen rupture, irrespectively of its cause, is called *defloration*. Delicate hymens, as well as thicker semilunar or ring-shaped hymens with a rigid orifice, tend to be ruptured relatively easily. Also, significant role is played by penile diameter. Sharper thrust of the penis and greater penile thickness increase the chance of hymen rupture and hemorrhage. Hemorrhage can be absent, if hymenal orifice is dilated by gradual, slowly defloration. There have been reports of cases of preserved hymen in 14-year-old prostitutes. Defloration is usually accompanied by pain sensation and hemorrhage. However, in case of rupture of a poorly vascularized zone of the hymen, visible hemorrhage can be absent. In approximately 50% of girls, the pain occurring at the defloration in beginning of sexual intercourse is insignificant or absent. Sometimes, the first sexual intercourse is not accompanied by any pain or bleeding.

Absence of pain or hemorrhage during the wedding night should not be considered evidence of lost virginity. Tight fibrous hymen can be a significant obstacle to sexual intercourse.

Clitoris (the Latin word "clitorido" means "I tickle"; the corresponding obsolete Russian term means "producing lust").

The clitoris develops from the genital tubercle; its embryonic development and morphology correspond to those of a part of penis.

The clitoris consists of the head (Latin: *glans clitoridis*), the body (Latin: *corpus clitoridis*), two cavernous bodies (Latin: *corpora cavernosa clitoridis*) and two pedis (Latin: crura clitoridis). In contrast to the penis, the clitoris does not contain the spongious part of urethra and does not fulfill the urine evacuation function **(Fig. 3.3)**.

The clitoris is located immediately under to mons pubis, above the vaginal vestibule; its head protrudes into the anterior corner or the vulvar slit in the form of a small cone-shaped tubercle. In many cases, the clitoris is hardly detectable during visual examination, and palpation identifies only a soft body whose size corresponds to the size of a millet seed; also, the clitoris can resemble a soft papilla. During sexual arousal, protruding clitoris becomes easily palpable.

The clitoris consists of two erectile cavernous bodies morphologically similar to penile *corpora cavernosa (Latin)*, as well as the head supplied by the dorsal neurovascular bundle. Externally, the cavernous bodies are covered by a dense fibrous tissue *tunica albuginea* (Latin), a sheath for cavernous bodies; this coat plays an important role during both clitoral and penile erection.

The *radix, body and glans* of the clitoris are distinguished; the clitoral head is covered by a double-layered fold of delicate skin called the *capuche*. Sometimes, increased density of the clitoral head can suppress the sensitivity of the clitoris.

Clitoral crura measuring about 1 cm in thickness originate from the ipsilateral ascending ramus of the ischial bone and descending ramus of the pubic bone and border the vaginal orifice on both sides together with *musculus bulbocavernosus* and *musculus ischiocavernosus*; thus, they are located deep inside the pudendal labia (Fig. 3.3).

Clitoral crura are descending and fusing together below of the pubic symphysis, to form the corpus clitoridis, covered by fascia. Glans clitoris, located on the top of corpus (column) clitoris, which is the more extremely sensitive erogenous area.

There is a wide variation in clitoral dimensions depending on the degree of its erection during arousal; they can reach to 3 cm. Clitoral head diameter is 2 to 10 mm.

Clitoral dimensions can increase in case of massive treatment with androgens, as well as in case of adrenal hyperfunction, disorders of sexual development (hermaphroditism) or constitutional virilism.

Obviously, there is no correlation between clitoral dimensions and female erotic reaction to clitoral stimulation. In approximately 40% of women, sexual arousal is accompanied by significant enlargement of the clitoral volume due to intensive inflow of arterial blood flow and comparatively smaller outflow of venous blood. Simultaneously, clitoris hardening and erection are contributed to by contraction of muscular fibers. As a result of clitoral erection, the glans clitoris approaches to the vaginal orifice.

Vascularization. The arteries supplying the clitoral (or penile) cavernous bodies (corpora cavernosa) have thick muscular layer and wide size. The arteries divide into several large branches making their way within the cavernous tissue septa. In case of the quiescent state of the clitoris (or penis), their course is spiral; this is why they are called *arteria helycinaea.* Internal coat of these arteries contains zones of increased thickness consisting of smooth muscle cell (myocyte) bundles and collagen fibers. During contraction of the artery walls, these zones of increased thickness act as valves to close the vessel. The veins also have thick walls and well-expressed muscular layer. The blood-filled cavities of cavernous bodies located between the arteries and veins have very thin walls lined with endothelium. The blood is drained from these cavities into deep veins via small arteriolo-venular anastomoses (connections). These anastomoses act as valves or shunts, because, during erection, their walls contract and close their lumen, thus preventing the blood outflow from the cavities and contributing to filling of the clitoral cavernous bodies [167-170].

Innervation. Clitoris is innervated by nerves of *plexus pudendalis* and *plexus hypogastricus caudalis*; sensory fibers pass within *n. dorsalis clitoridis* and *n. pudendalis* and finally reach the S1-S5 segments.

Summary

Innervation of female genital organs is ensured by *sympathetic* and *parasympathetic* parts of *autonomous nervous system*, as well as the somatosensory parts of the spinal nerves (**Fig. 3.2**).

The sympathetic fibers that innervate genital organs originate from aortic and celiac plexuses, pass downward and form the upper hypogastric plexus (*plexus hypogastricus superior*) at the level of the V lumbar vertebra (L-5). The upper hypogastric plexus gives off the fibers that form the right and the left lower hypogastric plexuses (*plexus hypogastricus sinister et dexter inferior*). The fibers from these plexuses pass to the large uterovaginal plexus (*plexus uterovaginalis*) and pelvic plexus (*plexus pelvicus*).

The uterovaginal plexus is located in the parametrial adipose tissue lateral and posterior of the uterus, at the level of the internal orifice of the uterine cervix and the cervical canal. This plexus is contributed to by branches of the *pelvic nerve* (n. pelvicus) belonging to the parasympathetic part of the autonomous nervous system. The sympathetic and parasympathetic fibers given off by the uterovaginal plexus innervate the vagina, the uterus, intramural parts of the uterine tubes and the urinary bladder.

The ovaries are innervated by the sympathetic and parasympathetic nerves given off by the ovarian plexus (plexus ovaricus).

External genital organs and the pelvic floor are mainly innervated by the *pudendal nerve (n. pudendus)* (**Fig. 3.3**).

1. <u>Somatic motor and sensory innervation.</u> Pudendal nerve (nervus pudendus) contains motor and sensory fibers innervating penis and clitoris. The motor neurons giving off the nerve fibers contributing to the pudendal nerve originating from the sacral plexus are located at the level of the S2-S4 segments. Sensory fibers reach the same level of the sacral spine.

The pudendal nerve has three branches. The first of these branches – the inferior rectal nerve – innervates the external anal sphincter. The second branch – the perineal nerve – ensures the innervation of the external urethral sphincter, bulbocavernous and ischiocavernous muscles, as well as other perineal muscles, perineal skin, scrotum (in males) and pudendal labia minora (in females). The third branch – the dorsal (sensory) penile (or clitoral) nerve, innervate the extremely sensitive for erotogenous stimuli, column and glans clitoridis.

2. <u>Parasympathetic innervation.</u> The bodies of the neurons that form the parasympathetic nerves are located in the sacral region of the spine. The preganglionic fibers pass within the ventral roots of spinal nerves S-2 / S-4 and the

cauda equina and form the pelvic nerves running from the lower hypogastric plexus or the pelvic plexus. The postganglionic fibers running from these plexuses innervate the erectile tissues of penis or clitoris, urethral smooth muscles (seminal vesicles and prostate - in males), as well as vagina and urethra (in females). Also, these nerves innervate the blood vessels that supply the pelvic organs and tissues related to genital functions.

3. Sympathetic innervation is ensured by the neurons located in the lateral horns of the lower thoracic and upper lumbar regions of the spinal cord. The preganglionic fibers leave the spinal cord within the ventral roots at the level of the T11-T12 segments and reach the sympathetic trunk, lower mesenteric plexus and upper hypogastric plexus. Postganglionic fibers pass within the hypogastric nerves and innervate the same organs and tissues as those innervated by the parasympathetic nerves [14, 16].

52

Fig. 3.1. Female Sexual Organ`s Anatomy

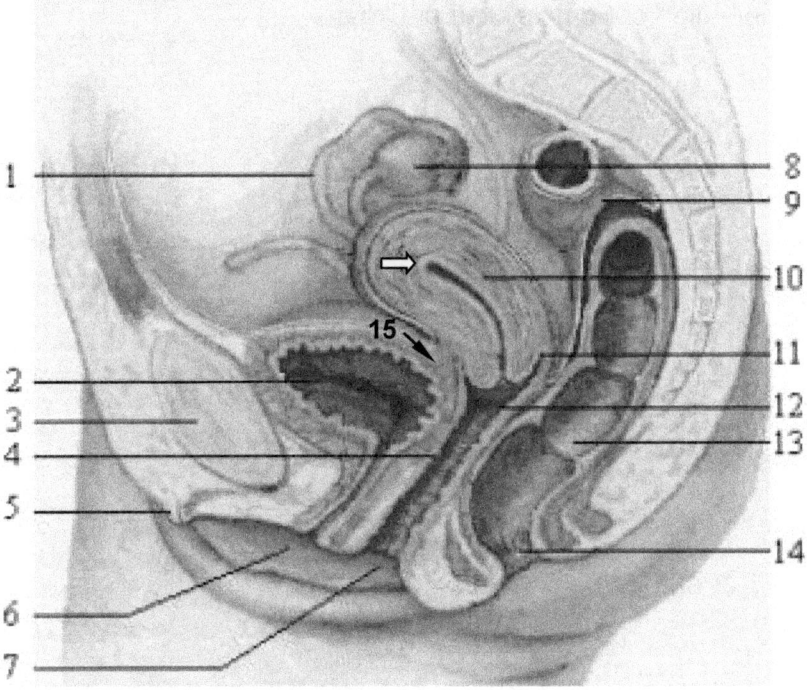

1. Tuba Uterina (Fallopian tube)
2. Urinary Bladder
3. Symphysis
4. Vagina
5. Urethra
6. Vestibulum Vaginae, terminated of Labia Minora
7. Vaginal Orifice
8. Ovarium
9. Colon descendens
10. Myometrium, white arrow – intrauterine cavity
11. Fornix vaginae posterior
12. Cervix
13. Rectum
14. Anus
15. Fornix vaginae anterior, **G-spot**

Fig. 3.2. Sympathetic and parasympathetic (autonomic) innervation of Female Sexual Organs

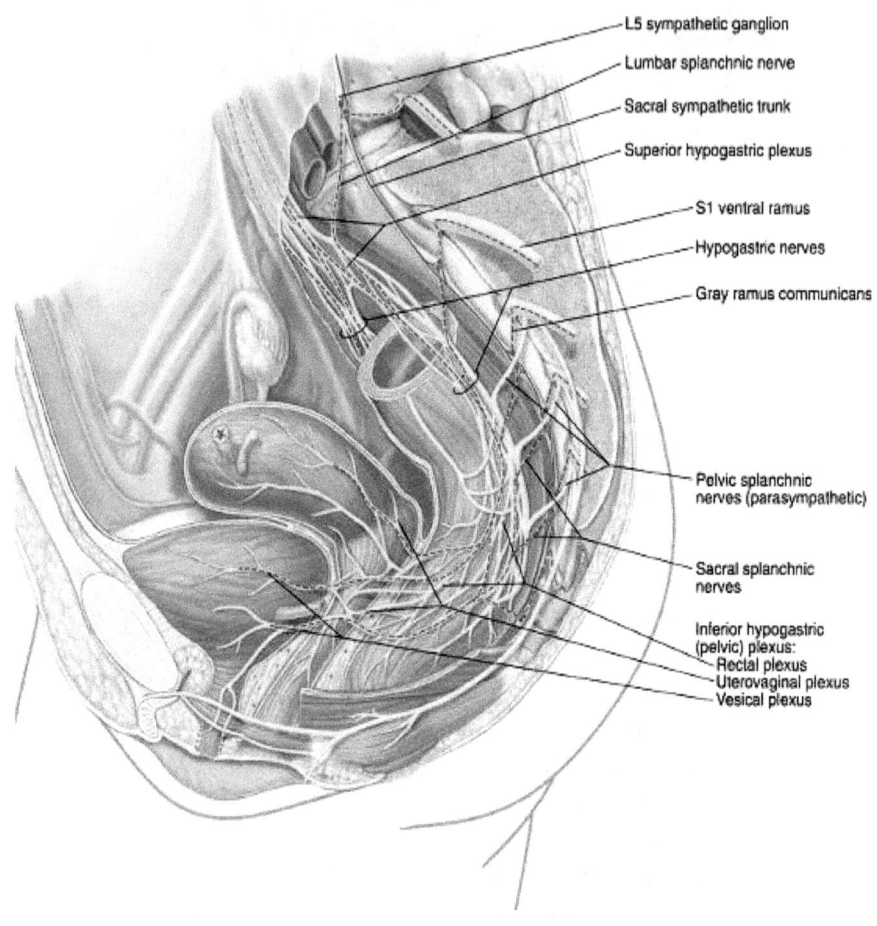

L5 sympathetic ganglion

Lumbar splanchnic nerve

Sacral sympathetic trunk

Superior hypogastric plexus

S1 ventral ramus

Hypogastric nerves

Gray ramus communicans

Pelvic splanchnic nerves (parasympathetic)

Sacral splanchnic nerves

Inferior hypogastric (pelvic) plexus:
Rectal plexus
Uterovaginal plexus
Vesical plexus

Fig. 3.3. Anatomy of Female External Organs.
Somatosensor innervation and vascularization.

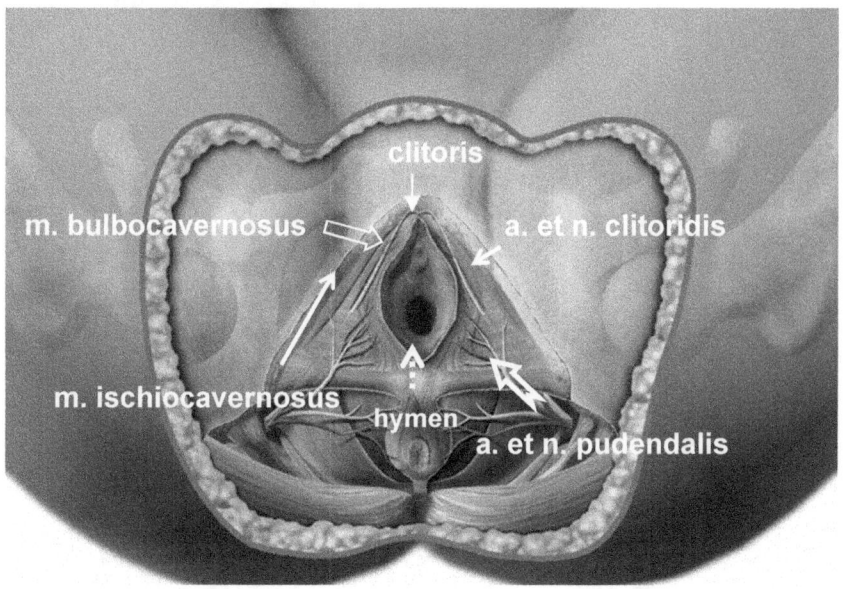

Nude-with-Snail-Breasts

Salvador Dali. Engraving in Giacomo Casanova's novels

Someday, when the study of anatomy will advance, it should be possible to link behavior of person and his passion.

marquis Donatien Alfons François de Sade, 1774

4. Physiology of sexual response: desire, arousal, orgasm

Human sexual response comprises several phases, typically distinguished as: *desire, arousal, and orgasm.* The neurophysiologic systems underlying these components are not well understood.

Sexual response is regulated through somatic and autonomic, peripheral and central nervous system pathways, and these pathways are modulated by steroid and peptide hormones.

Disruption of hormonal, neural, or vascular integrity is likely to interfere with normal sexual functioning, although psychological and relationship factors play important roles as well [33, 138].

Russian academician *Ivan Petrovich Pavlov* in experimental studying of Central Nervous neurophysiology (IP Pavlov. Complete Works. Leningrad, Vol. USSR Academy of Sciences, 1951, Volume IV, p. 25-26), had wrote, "The so-called Sexual, Digestive and Defensive instincts are complex of *unconscious (involuntary)* reflexes " [10].

For realize this program there are central mechanisms - cortex, subcortical neurons of the brain, the limbic system; and peripheral mechanisms - the reticular formation of the brain and spinal cord (see Figure 4.1), as well as erotic sensitivity of genitals and erogenous zones of the body.

Various models of sexual arousal and response have been proposed over the past century. In their description of the "sexual response cycle," *William Howel Masters* and *Virginia Eshelman Johnson* [25, 26] distinguished the sequential phases of sexual excitement, plateau, orgasm, and resolution, which corresponded to specific genital changes including: increased blood flow, the muscular contractions of orgasm, and the deactivation that follows. The model's strong focus on genital response and the semantic problem of using discrete labels for a physiologically continuous process were significant limitations.

Helen Singer Kaplan's model of sexual response, incorporating the three components: of desire, excitement, and orgasm, showed the interdependence among the response phases. For example, problems with orgasm could result from insufficient arousal; or problems with arousal might be seated in the desire phase. This triphasic model has had strong appeal because its components coincide with problems typically encountered by clinicians: lack of interest in sex, inability to become aroused - get an erection or show vaginal lubrication; or difficulty with orgasm - premature ejaculation, delayed orgasm, or anorgasmia [26].

Among the challenges for any model is that of identifying the underlying physiologic mechanisms of desire, arousal, and orgasm. In some instances, identification of supporting anatomic structures and neurophysiologic events has been quite successful, with respect to the local mechanisms involved in penile erection and vaginal lubrication. In others, the relationships are elusive; for example, a comprehensive understanding of the physiology underlying sexual desire and orgasmic response does not yet exist.

Neurobiological systems are involved in sexual response in 3 ways:

1) Physiologic input systems ensure sexual readiness (arousability) or induce sexual arousal itself. These systems convey information about general environmental conditions or context (is it the right time, the right place, the right mate?) or transmit specific sensory stimulation from a potential mate. In humans, the role of these systems is typically subtle and varies substantially between the sexes and among individuals.

2) Spinal and brain systems mediate sexual arousal and feelings. In humans, the relevant mechanisms and structures are only now being elucidated through positron emission tomography (PET) and magnetic resonance imaging (MRI) studies.

3) Finally, physiologic response systems are involved in the internal (autonomic) and external (somatic)

responses necessary for preparing and executing a sexual response.

The Role of Hormones in Women's Sexual Response

Arousability in most nonhuman mammals is largely under the control of the gonadal hormones. These hormones, produced by the ovaries and testes, are secreted in response to stimulation originating in the brain and mediated by the pituitary gland and its gonadotropic hormones. In sexually mature females, the secretion of gonadal hormones is sequential, with estrogen dominating during the first half of the reproductive cycle and progesterone during the second half. In males, the secretion of pituitary gonadotropins is tonic rather than cyclic, and the production of androgens is fairly constant over long periods.

The role of gonadal hormones in human female sexual response is less clear. In most female primates, ovarian hormones influence but do not control sexual behavior, which may be expressed even when gonadal hormones are minimal. In women, attempts to correlate desire, arousability, and arousal with different phases of the menstrual cycle, during which different hormones dominate, have met with only partial success. In women, both estrogens and androgens may work together to enhance sexual arousal and response. Variability in sexual interest in women, probably even more than in men, is likely to be contextual and partner-based, and relatively independent of biological control systems.

With aging and menopause, women may experience a number of sex-related problems that are directly or indirectly associated with decreased estrogen and androgen production. Sexual desire may be lessened and subjective and physiological aspects of sexual arousal may be affected, the latter because of reduced genital blood flow. Vaginal dryness and consequent discomfort or pain may also interfere with enjoyable intercourse.

While the specific mechanism through which gonadal hormones might facilitate sexual desire and arousal is unknown, they are probably active at multiple levels. For example, they may "prime" structures in the brain, thereby lowering the threshold to activation in the presence of sexually relevant stimuli. They may also work on spinal and peripheral neural systems: the rise of gonadal hormones during and after puberty may be responsible for "eroticizing" certain types of sensory stimulation—perhaps by transforming somatic sensory stimulation (such as genital touching) into autonomic information. Autonomic activation is generally associated with emotional responding and is necessary for feelings of excitement and arousal. Finally, these hormones may affect local mechanisms of penile and vaginal vascular competence. Most important, however, is the realization that sexual response is multifactorial, depending on complex interactions of physiologic, psychological, and relationship factors.

Central Mechanisms of Sexual Motivation and Arousal

Distinctions between the desire, cognitive-emotional, arousal, and response aspects of sexuality become blurred at the level of the central nervous system. Even in relatively simple animal models, the interaction of a number of structures is essential for sexual response, as sensory, information-processing, motivational, and motor elements of sexual response are integrated to generate a purposeful action. Furthermore, these structures may themselves be under the influence of modulators; specifically, structures involved in the control of sexual arousal and behavior are often sensitive to the presence of circulating steroid (gonadal and adrenal) hormones.

The *medial preoptic area* (MPOA) of hypothalamus involved in the regulation of sexual behavior in females, but its role is to inhibit sexual receptivity. The important brain structure responsible for activating sexual behavior in females appears to be the *ventromedial nucleus* (VMN) of the hypothalamus. The ventromedial nucleus may act by increasing the connection between sexual sensory stimuli and autonomic/ behavioral output.

The extent to which the preceding findings apply to humans is only now being clarified through research using MRI and PET. Preliminary studies suggest that some of the neural activation during sexual arousal may be shifted from lower brain centers medial preoptic area and ventromedial nucleus (MPOA and VMN) to higher brain centers in men and women.

This pattern is not surprising in view of the fact that sexual response in humans depends more heavily on contextual factors such as the relationship with the sexual partner, social behavior, attitudes, beliefs, and moral codes. In men, changes have been noted in the ventral *tegmental* area, a *midbrain-forebrain* region involved in mediating pleasure and reward. Concomitant changes in the frontal, occipital, and temporal lobes have also been noted. Generally, these brain regions interpret sensory stimuli and evaluate, choose, and execute motor/behavioral responses. Activity changes in hypothalamic and *amygdala* areas have also been noted in men. In women, many of these same structures appear to undergo change during arousal and orgasm. However, there appears to be less activation of hypothalamic and thalamic regions in women; this may explain the apparent differences between men and women in the response to visual erotic stimuli.

Neurochemistry of Sexual Response

Given widespread cerebral involvement in sexual response, it is not surprising that multiple neurotransmitter systems appear to contribute to it. However, exactly how and where various transmitters are operative, particularly in humans, are questions still under investigation.

In male rats, moderate doses of **dopamine** (D_1/D_2) agonists injected into the *ventromedial nucleus* promote erections and copulation. Furthermore, dopamine in the *nucleus accumbens* has been associated with many reward-related behaviors, including copulation.

Injection of the **nitric oxide** (NO) precursor L-arginine facilitates dopamine release in the ventromedial nucleus, suggesting a role there for NO as well.

Cholinergic nicotinic receptor influences on sexual behavior also appear to be mediated by the ventromedial nucleus.

Oxytocinergic neurons in the paraventricular nucleus (PVN) of the hypothalamus may further mediate erectile response, as do both ascending and descending **serotoninergic** systems in the brainstem.

The ascending (cerebral) serotonergic system generally exerts an inhibitory role on rat male sexual activity. However, the fact that different 5-HT receptor subtypes have different effects on sexual function may be related to serotoninergic involvement at different levels of the central nervous system— *forebrain, brainstem, spinal cord, and autonomic ganglia.*

Finally, alteration of **γ-aminobutyric acid** (GABA)ergic neurotransmission affects sexual behavior in rats. Both $GABA_A$ and $GABA_B$ agonists inhibit sexual behavior, whereas antagonists, at least when injected directly into the ventromedial nucleus, have prosexual effects.

Specific neurotransmitter involvement in females - both the ventromedial nucleus and paraventricular nucleus appear to have major roles in female sexual function. Furthermore, D_1 receptor activation — as well as oxytocin, GABA, and opioid receptor activation — in various hypothalamic centers can **increase lordosis (female receptive position) in rats.**

Likewise, serotonergic systems in the brainstem and upper spinal cord probably mediate orgasmic response in females. Given that drugs affecting serotonergic activity (eg, selective serotonin reuptake inhibitors) and dopaminergic activity (eg, antipsychotics) often interfere with women's orgasm, these systems are strongly implicated in human female sexual response.

In addition, drugs increasing GABA activity (eg, benzodiazepines) result in less satisfying orgasms for women.

Whereas drugs affecting nitrergic (involving NO), adrenergic, and cholinergic systems appear to have minimal effects on human female sexual response.

Peripheral Autonomic and Somatic (Motor) Responses

The somatic sensoromotor nervous system responds to information about the environment - visual, auditory, tactile, etc., and innervates striate muscles involved in *voluntary* motor responses. In contrast, the autonomic nervous system (ANS) is involved primarily in the control of *involuntary* internal smooth muscle responses, including *erection* and *vaginal lubrication.*

Sexual arousal and response require the activation of both systems, and their integrated functioning is complex.

During sexual arousal the autonomic nervous system (ANS) is activated via somatosensory stimulation to prepare the organism for sexual behavior. Activation of the autonomic nervous system is responsible for mediating extragenital smooth muscle changes, which are similar across the sexes, such as increased blood pressure, transient increases in heart rate, vasocongestion in the breast and pelvic regions, and, ultimately, an overall increase in muscle tension.

Genital changes, although different, tend to follow parallel courses in men and women.

Mechanisms of Erection

Both divisions of the autonomic nervous system, sympathetic and parasympathetic, are involved in arousal and activation of the genitals. The traditional functional classification of these systems (a homeostatic or regulatory role for the parasympathetic component and an emergency/arousal role for the sympathetic component) does not necessarily extend to activation of the genitals.

Thus, the parasympathetic and sympathetic components of the autonomic nervous system both appear to contribute to sexual excitement, clitoral (penile in men) erection.

Stimulation of parasympathetic fibers of the pelvic nerve arising from the sacral area of the spinal cord can generate clitoral erection. Recent studies, however, suggest a possible role for the sympathetic nervous system in erection as well, since blockage of this system produces clitoral (penile in men) engorgement and erection.

The autonomic nervous system influences erectile tissue through changes in the dynamics of blood flow of the pudendal arteries. Erection is the result of increased arterial flow through vasodilatation and shunting of the arterial blood away from immediate venal flow into the cavernous spaces of the clitoris (penis). This increase in arterial flow initially occurs without an increase in blood pressure and therefore is probably the result of relaxation of the smooth muscle of the arterial walls. When full erection occurs and intracavernosal pressure is increased, small blood vessels are compressed against the relatively unyielding walls of the tunica albuginea, and the resulting blockage decreases venal outflow.

Mechanisms of Vaginal Lubrication and Female Orgasm

In women (as in men), sympathetic, parasympathetic, and somatic pathways innervating the genital region, particularly the vagina and clitoris — mediate the response to sexual stimulation.

Sympathetic and parasympathetic nerves connect via the pelvic and pudendal nerves, and their stimulation can increase blood flow to and affect smooth muscle tone in the vagina. Somatic pathways are responsible for controlling striate muscles around the vaginal opening and in the pelvic and abdominal areas.

During sexual arousal, vaginal smooth muscle shows a gradual increase in tone. In addition, autonomic input stimulates blood flow to the vagina through vasodilatation; increased capillary flow, and the engorgement of the lining of the vaginal wall as well as the labia and clitoris with blood, stimulate vaginal lubrication. At the peak of sexual arousal all the capillaries are open and the flow is maximal. As with erectile response in men, α-adrenergic antagonists such as phentolamine cause vasodilatation and increased vaginal blood flow in women, suggesting that decreased sympathetic tone, along with concomitant increases in parasympathetic activity, mediates this response.

The trigger for orgasm probably involves accumulating afferent input. Undoubtedly both pelvic and genital structures: clitoris, uterus, cervix, etc., contribute to the overall experience of orgasm in women.

Clearly, the clitoris and, possibly, the periurethral glans (area below the clitoris surrounding the urethra) are homologous to the penis and are the locus of orgasm for most women. Clitoral, vaginal, and cervical self-stimulation differentially activated regions of the sensory cortex, but all were clustered in the medial paracentral lobule.

Because the perineal (groin) region is also stimulated incidentally during the clitoral, vaginal, and cervical self-stimulation, its corresponding sensory cortical region—i.e., the dorsal convexity of the paracentral lobule, immediately lateral to the midline—was also activated.

The present findings may help to resolve a discrepancy in the literature that claims that the location of the genital sensory cortical representation is on the dorsolateral paracentral lobule, rather than the medial paracentral lobule. That is, based on the present findings, the discrepancy in the literature may be due to the responses to indirect stimulation of the perineal (groin) region rather than to adequate stimulation of the genitals *per se*.

It is likely that the clitoris is indirectly stimulated by self-stimulation of the cervix or vagina. Under the conditions of the present study, it is not possible to discern whether the overlap among regions of the sensory cortex activated in response to self-stimulation of each of these three genital regions is due to true overlap of the brain regions that would be activated by "pure" stimulation of each of these three genital regions separately, or whether the overlap is due to incidental stimulation of one genital region (e.g., vagina) during self-stimulation of a different genital region (e.g., cervix). What is clear, however, is that the sensory cortical regions activated by each of these three genital regions are to some extent separable and distinct. Unexpectedly, nipple/breast self-stimulation activated not only the (expected) thoracic sensory homuncular region, but also the region of the paracentral lobule that overlaps with the region activated by clitoral, vaginal, or cervical self-stimulation. This finding is consistent with many women's reports that nipple/breast stimulation is erotogenic and can elicit orgasms (and personal communication).

The present finding of convergence between nipple and genital input in the genital sensory cortex is supported by an intriguing observation by Penfield and Rasmussen.

The ability of nipple stimulation to activate genital sensory cortex could have an indirect basis. Thus, nipple/breast self-stimulation-induced oxytocin secretion could stimulate uterine contractions that in turn generate afferent activity that projects to the paracentral lobule. However, it is also possible that nipple/breast and genital sensory activity converge directly not only on oxytocinergic neurons of the hypothalamic paraventricular nucleus, but also on paracentral lobule neurons of the genital sensory cortex.

The cerebellum activation observed in the present study during vaginal and cervical self-stimulation is a common observation during genital stimulation, especially during orgasm. It is likely that it is involved in controlling muscle tension during genital stimulation. Two other brain regions that were seen to be activated in the present study are the supplementary motor area and SII (secondary somatosensory cortex). Other brain regions activated more variably were thalamus, frontal, and parietal cortices.

Regarding the observation of bilateral activation of the hand representation in sensory cortex in response to unilateral hand-applied self-stimulation that was noted in the Results section for clitoral and vaginal self-stimulation, it is likely that the sensory stimulation emanating from that single hand, by utilizing the corpus callosum, generates contra- as well as ipsi-lateral activation of the hand representation in sensory cortex. This observation is supported by substantial evidence in the literature of bilateral sensory cortical response to unilateral hand stimulation. A more curious observation was the activation of the hand representation in sensory cortex during investigator-applied toe stimulation. One speculation to account for this observation is that subtle muscle-induced contractions of the hand in response to toe stimulation (a compensatory response preparatory to breaking the fall in the "stumble" response) activates the hand representation area in the sensory cortex, although no obvious hand movement was observed. Another possibility is that the response is among the class of atypical forms of referred sensation.

The present findings provide evidence that, rather than vaginal stimulation being just an indirect means of stimulating the clitoris, vaginal and cervical stimulation *per se* activate specific sensory cortical regions that are distinct from the clitoral sensory projection. These differential routes of entry into the brain are undoubtedly of significance in activating the diverse and differential consequences of clitoral, vaginal, or cervical stimulation; they include differential physiological effects (e.g., on prolactin secretion, analgesia, and blood pressure reactivity to stress) and differential behavioral effects (e.g., on orgasm, sexual satisfaction, and intimate relationship quality).

While the present study mapped the primary sensory field of genital input to the sensory cortex, it would be of interest in future studies to extend this analysis to brain fields beyond the sensory cortex that are activated when genital stimulation is perceived as "erotic" vs. when it is perceived as "just pressure".

As in men, peripherally mediated genital and orgasmic responses are under the control of brain and spinal mechanisms; alteration of central monoaminergic systems frequently interferes with orgasmic response in women. Furthermore, since ovarian hormones contribute to vaginal tissue response, female genital responsitivity is influenced by hormonal changes that occur with aging and various pathologies.

There is ongoing debate regarding the function of orgasm in women. Hypotheses include preparation of the uterus for impregnation, facilitated transport of sperm toward the uterus, or even dissipation of vasocongestion in the vaginal region.

Given these sex differences in structure and function related to orgasm as well as the brain structures involved, the mechanisms of orgasm may be sufficiently dissimilar in women and men so that the experience of orgasm is different as well.

Furthermore, 15% to 42% of women may experience multiple orgasms in rapid succession. In contrast, multiple orgasm in men is still viewed as "case study" material, although recent research suggests that a subpopulation of men may be capable of achieving multiple orgasms.

Conclusions

Sexual response involves interacting systems of physiologic input, maintaining arousability and producing arousal by transmitting general information about external conditions as well as specific sensory stimuli; spinal and cerebral input, inducing a state of central activation and arousal; and physiologic response, both autonomic and somatic, preparing and executing the climactic phase of the sexual response cycle.

Arousability is influenced by the gonadal hormones, but variations in sexual desire are more likely to be due to psychosocial than to biological factors. Responsibility for mediating arousal appears to be partly shifted from lower (limbic) to higher brain centers, and some of the brain structures that regulate arousal may be activated to differing degrees in men and women.

Somatic, sympathetic, and parasympathetic neural pathways are involved in the mechanisms of arousal and orgasm. Central regulation of sexual response primarily involves serotonergic and dopaminergic pathways, although several other neurotransmitter systems also take part.

Adrenergic, cholinergic, and nitrergic mechanisms mediate the peripheral vascular responses of penile erection and vaginal lubrication.

In both men and women, decreased sympathetic tone and increased parasympathetic activity responsible for increased genital blood flow during sexual arousal, and orgasm may be triggered by a reflex response to accumulating afferent input.

However, local differences in structure and function, together with differences in the relative involvement of some brain centers, suggest that there are differences, as well as common elements, in the way the cycle of arousal and orgasm is subjectively experienced by men and by women.

Salvador Dali. "**Oysters-and-Nude**"

In human nature is a secret inclination and desire to love others.

Francis Bacon, 1621

5. Female sexual disorders and treatment

Female sexual dysfunctions include:

➢ disorders of sexual desire (libido),
➢ disorders of sexual arousal – frigidity,
➢ pain or discomfort during sexual intercourse – dyspareunia,
➢ decreased or absent orgasm - anorgasmia.

Epidemiologic studies by Laumann and Rosen (2009) showed that 43% of women had at least one of the sexual specified problems [47]. Most of interviewed women had disorders of desire (libido) and arousal; however, the detailed clinical examinations revealed the predominance of orgasmic disorders (decreased or absent orgasm - anorgasmia).

Sexual problems, resulting of various psychologic, and somatic factors, necessitate qualified etiopathogenetic therapy. Despite these facts, female sexual disorders attract much less attention as compared to male sexual disorders (Laumann and Rosen, 2009), and there are only a few studies dedicated to physiological and psychological aspects of female sexual dysfunctions. Current classifications take into consideration mainly the psychological or mental problems and have been developed on the basis of self-assessment questionnaires [36, 47, 69, 122, 129].

Women's sexual desire, arousal, and orgasm disorders have been traditionally conceptualized, studied, assessed, and often treated from a perspective which compartmentalizes them.

However, despite the common finding in recent population-based epidemiological studies in which low sexual desire is the most prevalent of concerns relating to women's sexual functioning, women's sexual complaints are rarely experienced as discreet entities.

The Fourth Diagnostic and Statistical Manual of Mental Disorders, 4th Edition (DSM-IV) categories pertaining to lack of desire, arousal, orgasm problems or to sexual pain, are not independent, and in clinical practice, classification is often based on the way in which complaints are presented. Studies find comorbidity between desire and arousal. In part, this may be because women express difficulties differentiating desire from subjective arousal. Also, some women report their experience of desire to precede arousal whereas for other women, desire appears to follow arousal, as women do not seem to follow one universal sexual response cycle. As such, there has been notable criticism of the DSM-IV-Text-Revised (TR) classification of sexual dysfunctions in women [34, 36, 44, 131-136].

Table 1. Diagnostic criteria for women's sexual dysfunctions according to the Diagnostic and Statistical Manual of Mental Disorders, 4th Edition, Text-Revised (DSM-IV-TR) and American Urological Association Foundation (AUAF) [34, 36].

Desire	
DSM-IV-TR	Hypoactive sexual desire disorder: Persistently or recurrently deficient (or absent) sexual fantasies and desire for sexual activity. The disturbance causes marked distress or interpersonal difficulty
AUAF	Sexual interest/desire disorder: Absent or diminished feelings of sexual interest or desire, absent sexual thoughts or fantasies, and a lack of responsive desire. Motivations (here defined as reasons/incentives) for attempting to become sexually aroused are scarce or absent. The lack of interest is considered to be beyond a normative lessening with life cycle and relationship duration.

Arousal	
DSM-IV-TR	Female sexual arousal disorder: Persistent or recurrent inability to attain, or maintain until completion of the sexual activity, an adequate lubrication-swelling response of sexual excitement. The disturbance causes marked distress or interpersonal difficulty. The sexual dysfunction in not due exclusively to the direct physiological effects of a substance (e.g., a drug of abuse, a medication) or a general medical condition.
AUAF	Subjective sexual arousal disorder: Absent or markedly diminished feelings of sexual arousal (sexual excitement or sexual pleasure) from any type of sexual stimulation. Vaginal lubrication or other signs of physical response still occur. Genital sexual arousal disorder: Absent or impaired genital sexual arousal. Self-report may include minimal vulval swelling or vaginal lubrication from any type of sexual stimulation and reduced sexual sensations from caressing genitalia. Subjective sexual excitement still occurs from nongenital sexual stimuli. Combined genital and subjective arousal disorder: Absent or markedly diminished feelings of sexual arousal (sexual excitement or sexual pleasure) from any type of sexual stimulation as well as complaints of absent or impaired genital sexual arousal (vulval swelling, lubrication).

Orgasm	
DSM-IV-TR	Female orgasmic disorder: Persistent or recurrent delay in, or absence of, orgasm following a normal sexual excitement phase. Women exhibit wide variability in the type or intensity of stimulation that triggers orgasm. The diagnosis of female orgasmic disorder should be based on the clinician's judgment that the woman's orgasmic capacity is less than would be reasonable for her age, sexual experience, and the adequacy of sexual stimulation she receives. The disturbance causes marked distress or interpersonal difficulty. The orgasmic dysfunction is not due exclusively to the direct physiological effects of a substance (e.g., a drug of abuse, a medication) or a general medical condition.
AUAF	Women's orgasmic disorder: Despite the self-report of high sexual arousal/excitement, there is either a lack of orgasm, markedly diminished intensity of orgasmic sensations, or marked delay of orgasm from any kind of stimulation.

➤ Definition of Hypoactive Sexual Desire Disorder (HSDD)

A critical look at existing definitions of female sexual dysfunctions is warranted given that they have a direct and profound impact on instrument development, epidemiological studies, treatment protocols, etc.

The DSM-IV-TR diagnosis of HSDD focuses on "persistently or recurrently deficient (or absent) sexual fantasies and desire for sexual activity" which causes marked distress or interpersonal difficulty (Table 1). This definition has been criticized as overpathologizing women on the basis that women themselves may not necessarily consider sexual fantasies and desire for sexual activity to be an index of their sexual desire and that some women may deliberately evoke fantasy as a way of boosting their sexual arousal. Moreover, in two large recent prevalence studies of older women, 70% reported desiring sex less than once a week but the majority (86–89%) were at least moderately to extremely emotionally sexually satisfied; and in the second study, the majority (71.2%) of women with low desire were happy with the relationship. There is also desynchrony between sexual satisfaction and frequency of sexual activity. Collectively, these studies suggest that women may experience a satisfying sexual life with a partner without the outright desire for sexual activity [34, 43, 47, 48, 62, 69, 122, 137].

Mounting evidence suggests that all sexual desire is responsive. There are a large number of cues that provoke sexual desire and sexual activity in women.

Engaging in sexual activity in the absence of an identifiable external trigger (e.g., "because the opportunity presented itself","because I was in the mood") was an unlikely reason women provided for having sex.

The sexual desire may be regarded as the result of an incentive (sexually competent stimulus) which activates the sexual system, of which the subjectively perceived desire is one of many components.

> ## Definition of Female Sexual Arousal Disorder (FSAD)

In the DSM-IV-TR, FSAD is defined as the "persistent or recurrent inability to attain, or to maintain until completion of the sexual activity, an adequate lubrication-swelling response of sexual excitement," coupled with "marked distress or interpersonal difficulty." In contrast to earlier editions of DSM, subjective sexual experience is not part of the definition.

There are a number of problems with these criteria. Lubrication problems are not necessarily distressing for women. In a closer exploration of personal (interpersonal) distress associated with lubrication problems, 31.2% of the sample complained of lubrication difficulties but only 7.3% reported "marked distress" about the relationship and 6.5% personal distress. Lack of lubrication is often a poor predictor of distress, except among postmenopausal women. In the clinical setting, complaints of "genital deadness" or absent/impaired subjective sexual arousal are far more common [34].

Another difficulty in making the diagnosis relates to the determination by the clinician that the women have received an adequate amount of sexual stimulation. The precise definition of "adequate" arousal may vary across women; for some, adequate sexual arousal involves physical as well as "psychological" and "situational" stimulation. Also, in the case of stimuli no longer being effective, is this indeed FSAD or a case of habituation? The clinician has to evaluate what is normal, based on age, life circumstances, and sexual experience. There is a great variety in the ease with which women can become sexually aroused and in which types of stimulation are required.

Women's genital response coincide with subjective experience and based on their appraisal of the situation.

➢ **Definition of Female Orgasmic Disorder (FOD)**

Women have difficulty perceiving genital changes associated with sexual arousal, and because women who report lack of orgasm may in fact be insufficiently sexually aroused, the distinction between *female sexual arousal* (FSAD) and *female orgasmic disorder* (FOD) is unclear. *Female orgasmic disorder* are defined as the persistent or recurrent delay in, or absence of, orgasm following a normal sexual excitement phase (although it does not stipulate whether it is normal subjective or physiological arousal being referred to). In cases where the clinician does not have access to a psychophysiological test, it cannot be established that her deficient orgasmic response occurs despite a normal sexual excitement phase, unless she reports feelings of sexual arousal.

The lubrication/swelling response depends on adequate sexual stimuli, degree of sexual arousal and desire.
Success of directed masturbation for anorgasmia suggests that lack of adequate sexual stimulation is an important etiological factor underlying lifelong, and probably also acquired, *female orgasmic disorder.* The problem *of anorgasmia or low orgasm* can be improved by adequate sexual stimulation. Sexual satisfaction is an important component of contemporary models of women's sexual response, and it is recognized to have personal and relational domains.

Etiological Factors in Desire and Arosual Disorders

From the perspective of Emotion Theory, sexual desire may be conceived of as being connected to other basic emotional systems like fear–anxiety, in that it is a highly adaptive response to an emotionally competent stimulus [52]. In this perspective, the subjective experience of desire may be the conscious awareness of the automatically generated bodily responses to the stimulus (i.e., arousal) that produces the sensation of "wanting". The subjectively experienced state of desire may thus be the final result of a complex interplay of driving and inhibiting forces. The biological factors mentioned below may hamper responses to "sexually competent stimuli".

The most important neurotransmitters involved in desire and subjective arousal identified so far are norepinephrine, dopamine, melanocortins, oxytocin, serotonin acting via 5 HT1A and 5 HT2C receptors—being prosexual, and prolactin, GABA and serotonin acting via other receptors—being inhibitory or negative. The actions of these substances are modified and influenced by the endocrine homeostasis provided by: follicle stimulating hormone (FSH), luteinizing hormone (LH), estrogen (E), progesterone (Pg), and testosterone (T). Extensive research has shown the impact of chronic medical illness, childbirth, and oral contraceptives on sexual desire and arousal.

Physical Examination

Gynecological examination is recommended for reasons of good medical care and for education and reassurance, and in the assessment of women with desire and arousal complaints to assess and rule out medical/physical contributors.

The investigation is also a setting to explore perceptions, beliefs, and attitudes about a woman's own anatomy and encourage a patient's positive approach to her genitals and body.

Gynecological examination should include an evaluation of the level of voluntary control of the pelvic floor muscles, pelvic floor muscle tonus, presence of vaginal wall prolapse, signs of vaginal atrophy, size of introitus, presence of discharge, or evidence of infection (acute or chronic), epithelial disorders, and/or pain. In some disease conditions (e.g., multiple sclerosis or pituitary disease), there may be specific symptoms of vulvar sensory loss, atrophy, or hypertrophy. The physical examination is particularly important for ruling out or identifying medical factors when concomitant complaints such as loss of sensitivity or sexual pain exist.

Laboratory Investigations. The possibility that laboratory testing will identify causes of sexual dysfunction is low. Estrogen deficiency is best detected by taking a history and performing a physical examination.

Hormonal measurements are indicated of women with: irregular menstrual patterns or amenorrhea, low arousal and hypoactive desire, in menopausal or hysterectomized women without a clear symptom history.

Most studies have failed to find a correlation between low sexual desire and serum T levels in women. In women with symptoms or signs of thyroid disease or hyperprolactinemia (galactorrhea, irregular menses, and/or infertility), diagnostic assays should be taken.

Estrogen and Sexual Response. The most abundant and potent estrogen before menopause is 17β-estradiol (or estradiol). Estriol and estrone are present at much lower levels and display less activity on estrogen receptors [146]. The primary source of estradiol is the granulosa cells of the ovaries. After menopause, estrogen is produced in extragonadal intracellular sites from dehydroepiandrosterone (DHEA), dehydroepiandrosterone sulphate (DHEAS), and androstenedione (A4). The major estrogen in serum of postmenopausal women is estriol which is not measured by clinically available assays.

Clinical studies have shown that an adequate estradiol level is important for maintaining vaginal lubrication and avoiding dyspareunia. Effects of menopause and hypoestrogenism on genital sensitivity are not well known, but low level of estradiol correlates with decreased sexual desire in postmenopausal women. Lack of estrogen has been found related to dyspareunia and vaginal dryness. However, many hypoestrogenic women do not have dyspareunia. When sufficiently sexually stimulated, low estrogen may be less important. An association between reduced estrogen and decreased sexual desire has been only described clinically. There is some limited evidence to affirm that there is an association between reduced vulvar sensitivity to pressure/touch and estrogen deficiency

Androgens and Sexual Response. The androgens include testosterone (T), dehydroepyandrosterone (DHEA), dehydroepyandrosterone-sulfat (DHEAS), androstendyone (A4), and 5 α-dihydrotestosterone (DHT). Of the androgenic steroids, T and DHT have the most potent biological activity. DHEA, DHEAS, and A4 are adrenal and ovarian precursor steroids that can be metabolized into T, DHT, and estrogen in peripheral tissues.

Androgens circulate in the body bound by a variety of proteins, including albumin, cortisol-binding globulin, α2-glycoprotein, and most importantly, sex hormone-binding globulin (SHBG). Androgens bound to SHBG are essentially not bioavailable; in contrast, androgens complexed to albumin are rather available because of their lower affinity. Androgen levels peak when women are in their 20s and drop gradually with age, so that women in their 40s have approximately half the level of circulating total T as women in their 20s. T levels do not decline consistently during or after menopause.

<u>Androgens are known to act on multiple tissue and receptor
sites throughout the body.</u>

In 2002, a Consensus Conference on androgens agreed that
androgen insufficiency in women with adequate estrogen
levels could lead to a diminished sense of well-being and
energy, fatigue, and decreased sexual desire! [172, 173]

More recent guidelines from the Endocrine Society
recommended against making a diagnosis of "androgen
insufficiency," because of the lack of a well-defined clinical
syndrome and normative data on T levels across the lifespan
that can be used to define such a disorder

Neurotransmitters and **steroid (sex) hormones** appear to
have a modulatory function on each other, and changes in
one system may have a dramatic effect on the other. For
instance, recent longitudinal data suggest that the transition
to menopause is strongly associated with new onset
depressed mood among women with no history of depression
[173].

Clinical data of new imaging techniques such as positron-
emission tomography, and functional magnetic resonance
imaging (MRI) have confirmed: the major neurochemical
systems involved in sexual behavior consist of pathways
involving neurotransmitters and hormones including
dopamine, serotonin, norepinephrine, prolactin, oxytocin,
melanocortins, and endogenous opioids. Some of these

substances are substantially influenced by gonadal steroid hormones and/or interact among each other.

Treatment of Low Desire

Before starting a therapy of sexual dysfunction, it is necessary to carry out a differential diagnosis between **sexual anesthesia, absence of sexual desire** and **vaginism** during physical and gynaecological examinations.

Vaginism (from the Latin term *"vagina"*) means involuntary convulsive contraction of the vaginal and pelvic floor muscles. Sometimes, vaginism involves convulsive contraction of femoral and/or abdominal wall muscles.

Etiology of vaginism is vary, usually occurring in young women after painful sexual intercourse, with previous severe painful traumatic injury in genital organs, painful defloration, pathologic delivery with vaginal trauma, vulvar ruptures, childhood perineal trauma. Vaginism can be caused by carelessly conducted and painful gynecological procedures, as well as insufficient anesthesia during gynecological procedures, vaginal examination, vaginal surgery.

The forms of vaginism specified above mainly results from psychological reaction due to previous genital trauma. Convulsive muscular contraction excluding the possibility of vaginal penetration may be occurring if sexual penetration causes to feel of woman severe pain, due to dry vagina, vulvovaginitis (inflammation of vagina), and damaged external genital organs. In these cases the muscular contraction represents a protective reflex response and called "pseudovaginism" or "dyspareunia" in contrast to vaginism [140-144].

It should be noted that a lot of gynecological diseases are associated with somatic changes, causing *vaginism* and *dyspareunia* (painful sexual intercourse); they include endometriosis, some variants of uterine myoma, congenital vaginal anomalies, vulvar scars after traumatic delivery and colpitis. So, sexual disorders curation (treatment) should be made only after exclusion of gynaecological diseases.

Hormonal Treatment of Low Desire. Testosteron (T) has been used in the treatment of low sexual desire since the 1930s; however, systematic study of it is only relatively recent. Studies in surgically menopausal estrogen-replete women who reported a decline in their desire for sex have found a benefit of T administered via a 300 μg/day patch, but no significant beneficial effect of either 150 μg/day or 450 μg/day compared with placebo. Similar effects were found among naturally menopausal, estrogen replete women.

The androgenic, progestogenic, and estrogenic synthetic hormone, tibolone, is available in Europe. Clinical studiies found a significant increase in scores on the Female Sexual Function Index (FSFI) following 24 weeks of use.

Testosteron therapy is effective for estrogen-replete naturally menopausal women, and marginally effective for premenopausal women, though it produces supraphysiological levels in the latter [69, 70, 153, 172].

Nonhormonal medications that have been investigated for low sexual desire have typically had a mechanism of action that was centrally acting. In nondepressed women with *hypoactive sexual desire disorders* (HSDD), the antidepressant buproprion, which blocks norepinephrine and dopamine reuptake, was found to significantly improve sexual arousal and orgasm, but not sexual desire. In women with selective serotonin reuptake inhibitor (SSRI)-associated mixed sexual symptoms, 4 weeks of treatment with the addition of bupropion led to a significant increase in self-reported feelings of desire and sexual activity.

The most recently investigated of the centrally acting agents for *hypoactive sexual desire disorders* (HSDD) has been flibanserin. Flibanserin's mechanism of action is not yet fully understood but it acts as a 5-HT1A serotonin receptor agonist and 5-HT2A serotonin receptor antagonist. At present, no peer-reviewed publications concerning efficacy of flibanserin on women's sexual desire are available, but results of flibanserin 100 mg taken daily were made public at the European Society of Sexual Medicine annual meeting in November 2009 and summarized at the Boehringer Ingelheim website. Of note, Boehringer Ingelheim categorizes *hypoactive sexual desire disorders* (HSDD) as a medical disease, but unclear is how it was established that the complaints were of medical etiology and not resulting from other sources. The primary endpoint was frequency of satisfying sexual events (SSE) following 24 weeks of treatment, or the number of sexual events (defined as sexual intercourse, oral sex, masturbation or genital stimulation by the partner) which were satisfying for the woman (i.e., gratifying, fulfilling, satisfactory and/or successful), irrespective of whether women had an orgasm or whether the event was satisfying for the partner.

Flibanserin everyday taking 100 mg had statistically significant improvements in their level of sexual desire as measured by the eDiary and a reduction in female sexual distress. Most adverse drug reactions in flibanserin 100 mg were mild to moderate and included dizziness, nausea, fatigue, somnolence and insomnia. Given that flibanserin's mechanism of action is not understood, flibanserin is the best treatment for women with absent or low sexual desire [157].

➢ **Treatment of Low Arousal Psychological Treatments**

Therapy of female sexual disorders should take into consideration the ethnic, cultural and ethic factors. Desire, arousal and orgasm are equally important aspects of sexual function. It is necessary to evaluate the degree of sexual stimulation in the healthy women who have no disorders of arousal, but can develop them, if affected by unfavorable

96

factors. In the women whose somatic status depends on the use of pharmacotherapy, arousal can increase in response to drugs stimulating libido.

The up-to-date approaches used in the medical practice: hormone replacement therapy; surgical correction of pelvic floor muscle failure, as well as genital organ prolapse; and methods of aesthetic surgery are able to improve the physical status of the patients. Appropriate methods of pathogenetic therapy and necessary surgical correction should be chosen by gynecologist.

Frigidity (Latin: frigiditas) – sexual coldness in sexually inert woman, they have not physiological manifestations of arousal, vaginal wetting (lubrication) during arousal or sexual intercourse.

Some authors excessively expand the range of conditions covered by the term frigidity by categorizing all women unable to reach vaginal orgasm.
According to statistics [108, 109, 115] approximately 60-90% of women may be considered as frigid. They had no erotic feelings or sexual pleasure caused of inability to reach orgasm for a long time.

Our data shows that frigidity is present in only 33% of the anorgasmic women have been married for at least one year. Her intelligent attitude toward sex is inconstant. Sex might her feeling for disgust, leave neutral or feel pleasure.

Fortunately, completely inert women are rarely encountered. And this is indeed a fortunate fact, bearing in mind the extreme complexity of treatment of this condition [117-120].

Frigidity can be *transient* or *permanent*; also, it can be noted since the beginning of sexual life - primary frigidity; or start at some point in sexual life in the women having orgasms previously - secondary frigidity [70, 77-81, 111].

Complete female frigidity - frigidity of woman in all cases of her personal contact with a man is one of the most complex clinical disorders, and its sex therapy is characterized by the most pessimistic prognosis. The prognosis is more favorable in a woman previously showing sexual response in certain situations, but currently frigid in relations with her husband. However, intensive psychotherapy is necessary even in cases of situational or partial frigidity. [112]

Frigidity psychodynamics is characterized by the following main manifestations. The woman develops a subconscious conflict interfering with her ability to get pleasure from physical contact with a man. As a rule, this form of sexual disorder is not associated with any particular conflict and presents itself as a syndrome. Manifestations of this disorder in a woman include the fears connected with the Oedipus complex; animosity toward men in general or any individual man in particular; the fear of being rejected in case of allowing herself to internally relax; bothering for "appropriateness" of her sexual behavior; feeling ashamed for her sexuality manifestations. In fact, the specific protective mechanisms characteristic of frigid women prevent them from expressing their sexual response. The woman either avoids obtaining adequate stimulation from a man or, in case of allowing her lover to stimulate her, develops perceptive protection preventing her from obtaining pleasure from this stimulation. She literally forbids herself to feel any erotic sensations. She subconsciously forbids any sexual response on her part and does not allow herself to falling into erotic sensations [35, 67, 82-86, 88, 112].

Tibolone treatment of frigid women resulted in a significant increase of erotic fantasy. Tibolone was associated with significant increases in sexual desire, and the frequency of arousability and of sexual fantasies. Vaginal lubrication was significantly improved on tibolone. In another study, tibolone showed similar benefits on sexual function, where women receiving tibolone had significantly higher sexual desire, sexual excitement, intercourse frequency, and vaginal dryness scores.

The main therapeutic approaches consists in the appropriate structuring of the woman's sexual situation so as to ensure her adequate response to stimulation on the background of her relaxed, placid and amorous state of mind. In this situation, she has to learn to restore her erotic feelings that have been inhibited by protective mechanisms for a long time [67, 82-86, 111].

Treatment methods of Anorgasmia

Anorgasmia is failure to experience an orgasm during sexual intercourse; according to gynecologist Helene Michel-Volfromm (1965), anorgasmia was present in 40% of French women.
Helene Stourzh (1962) had evaluated the quality of sexual life of 1500 women in Vienna, so one third of them never had experienced of orgasm, and 65% of the anorgasmic women had sexual intercourse with more than one male [110].

Some cases of increased sexual activity during certain age periods (e.g., puberty, perimenopause, etc.) have been reported in literature.

Hypersexuality is a pathological increase of sexual desire and orgasmic ability. Pathological increase in sexual desire is called *sexomania* or *nymphomania* (for females) and *satyriasis* (for males), based on the terms *nymphs* and *satyres* found in the Ancient Greek mythology. Pathological increase in orgasmic ability is called *hyperpotency*.

S. Freud (1905) described a female patient who was able to have 1 to 2 orgasms in the course of one sexual intercourse when in her usual mental state and 20 to 30 orgasms in the course of one night when in a hypomanic state; her hypomanic state was also accompanied by sharp increase in the sexual desire. In his another female patient who suffered a personality disorder complicating encephalitis, sharply increased level of sexual desire, but preserved level of orgasmic ability, were noted.

Hypersexuality is a polyetiological condition. One of its causes is a hypothalamic and/or limbic disorder resulting from previous neuroinfections (e.g., encephalitis), traumatic or vascular cerebral injuries, cerebral tumors or effects of certain narcotic agents. These cerebral regions are directly related to sexual function regulation.

An extensive review of the literature has been published following the previous 2003 International Consultation on Sexual Dysfunctions and at present, there are no significant new data on the subject except preliminary data showing significant heritability in orgasmic function either alone or with a partner [219].

Medications of Anorgasmia. One trial of sildenafil in female orgasmic disorder (FOD) who were recruited over 4 years across seven American treatment centers, those in the treatment group, had significantly therapeutic effect on sexual function quality [110].

Treatment of Anorgasmia. Methods of Sex - Therapy developed by Masters and Johnson [25] and Professor Helen Kaplan Singer [26] are the main methods of therapy *anorgasmia* remain directed masturbation in conjunction with sex education, anxiety reduction techniques are the main therapeutic tools for anorgasmia.

Professor *Helen Kaplan Singer* [25] proposed to use sexual exercises as a central element of the so-called **sex therapy** [26, 70, 76].

Masters and Johnson introduced the term *focusing on sensation* in sexologic practice. Sex-Therapy may to improve sexual function in 57% of women with sexual desire disorder. Traditional sex therapy techniques enhances quality of sexual life in 65% of 365 married couples [25].

As a rule, it is not necessary to force the achievement of orgasmic reaction in a frigid women at the initial stages of sex therapy. The therapy should be aimed at the general augmentation of the response. Paradoxically though it may seem, excessive concentration on orgasmic reaction usually becomes an impeding rather than accelerating factor in the course of sex therapy.

Most often, women complain of their inability to have an orgasm during sexual intercourse. This fact is quite understandable in view of their specific physiological and anatomical features. Experimental and clinical data shows that female orgasm is a reflex whose motor part involves sequential contraction of the muscles adjacent to the vaginal region notwithstanding the fact that the source of stimulation is located in the clitoral zone. From the anatomical viewpoint, therefore, sexual intercourse is neither direct nor optimal way of clitoral zone stimulation.

It is necessary to distinguish between the completely unresponsive (i.e., completely frigid) women and the women able to have erotic feelings, normal vaginal moistening (lubrication) and vasomotor hyperemia, but having various problems related to orgasm achievement.

Previously, both partial and complete unresponsiveness in females were covered by the term "frigidity" and regarded as one and the same disorder. This is an erroneous viewpoint, because these two syndromes, though having much in common, result from different mechanisms and require the use of different sexotherapeutic approaches for their treatment.

There is much controversy regarding the diagnosis of orgasmic dysfunction in view of the absence of generally recognized ranges or limits of normal manifestations of female orgasm. Orgasm, like any reflex, is characterized by definite parameters, dynamics and thresholds.

In her monograph "Sex therapy", professor of psychiatry Helen Kaplan Singer distinguishes the following variants of female sexual anesthesia (frigidity) [112]:
  o females with no previous orgasms,
  o females requiring intensive clitoral stimulation (i.e., the females who reach an orgasm by themselves and are "not interested" in a sex partner),
  o females requiring a direct clitoral stimulation during sexual intercourse and able to have an orgasm,

- females able to reach an orgasm during sexual intercourse, provided that preliminary prolonged and active stimulation of erogenous zones (especially, clitoris) has been used,
- females reaching maximum pleasure after short vaginal penetration during sexual intercourse,
- females needing fantasies and/or breast stimulation for reaching an orgasm.

Final results of pharmacotherapy and various relaxation procedures can be assessed on the basis of subjective arousal, feeling of pleasure and degree of satisfaction with sexual relations [97-103, 116, 121].

Approaches for solving this problem recommended by Professor Helen Kaplan Singer on the basis of her valuable empirical data. Sex therapy is indicated for all women who complain of low arousal and anorgasmia, including both complete and partial orgasmic disorders.

<u>The main principle for reaching an orgasm is simple: use a combination of increased stimulation and decreased sexual inhibition</u> [112].

Usually, self-stimulation carried out of female in either the presence or absence of a sex partner is found to be an extremely effective method for decreasing sexual inhibition.

Orgasm is not only final BLISS!

Recognition of this fact by a woman leads to resolution of many problems and significantly facilitates her focusing on erogenous feelings and obtaining sensual pleasures and thus, ultimately, results in the reaching of maximum pleasure.

Obviously, human beings are monogamous and tend to get married and stick with one partner for their whole life or transiently. Amorousness, which can be regarded as a human analog of attachment observed in animals, is characterized by the recruitment of quite predictable behavioral models and, beginning from the juvenile period, is associated with secretion of specific hormones acting on the brain cortex. The majority of people feel better and more confident when they are in love. Sex with own beloved one brings incommensurably more happiness than sexual activity with an unfamiliar partner.

However, falling in love is contrary to the nature of some people and other people have distorted amorous and sexual manifestations.
Thus, many males and females demonstrate better sexual functioning with the partners not having stabile emotional or other relations with them, or in conditions of frequent change of sex partners.

Any biological behavioral program, irrespective of its ultimate effects, acts as an integral part of the human sexual motivation. The acquired knowledge, sensations and experience play a significant role in the human behavior within the domain of amorous and sexual relations.

Though our knowledge in this field is insufficient, the accumulated clinical experience shows that sexual functioning of a healthy person who has no guilt feelings does not depend on whether he or she loves his/her partner or not. Humans are able to have active and normal sex life and get erotic pleasure even in the absence of strong romantic feelings, provided that their sexual relations with their partners are not destructive, and their partners do not provoke the feeling of disgust.
Love makes sexual relations infinitely joyful and human and causes them to bring genuine pleasure.

"Sex therapy is not able to plant the seeds of love.
However, the process of treatment often eliminates the barriers of feelings and expressions" [26]. *Helen Kaplan Singer*

Sometimes humans look for *deviant* ways to satisfy of sexual desire.

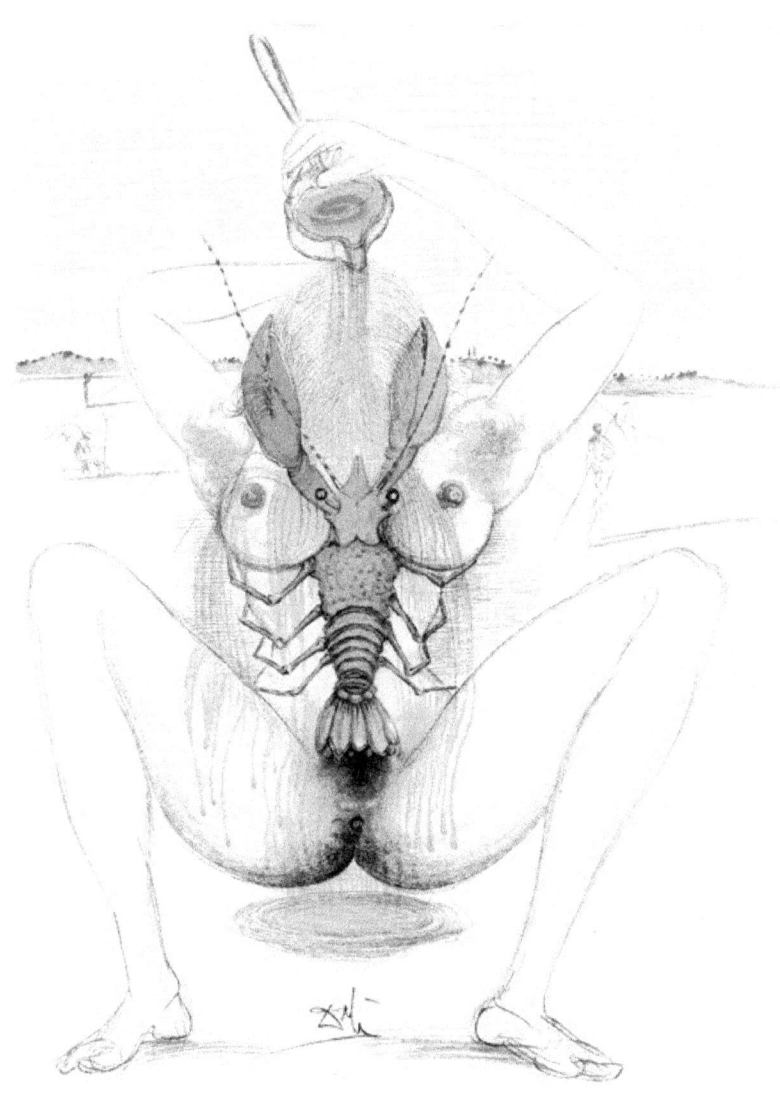

Salvador Dali. "Nude-and-Lobster"

Humans condemn passion, forgetting that inspiration lights his torch from their fire.

marquis Donatien Alfons François de Sade, 1772

6. Sexual deviations[1]

Sexual orientation is one of the attributes determining gender identity and, accordingly, reproductive behavior of the human being. In the XVIIth - XIXth centuries, as well as in the first half of the XXth century, deviations from the standard heterosexual orientation were considered as mental disorders of unclear etiology [11].

As a consequence, deviations from standard heterosexual orientation were addressed only by psychiatry and, partially, psychology.

This situation started to change in the XXth century when, as a result of development of endocrinology, genetics, embryology, neuroendocrinology and biology, effects of hormonal factors on human physiology and behavior were demonstrated deviations of nature.

Embryonal studies of French endocrinologist Alfred Jost (1976) revealed the mechanism of somatic sex differentiation of human embryo. According concept of Alfred Jost, human

---

[1] Sigmund Freud used this term to denote deviations from normal sexual relations.

In monograph, this term used for characterize the possible differences in, and multiplicity of, orientation, behavior and preferences. *Author's notes*

embryo (fetus) develops independently of genetic gender as female without initial virilization factors (without androgen). Androgen-induced influences have virilizing effects on the embryonal genitalia and forming male-type development.

The situation changed to adoption of a more tolerant attitude to patients with congenital sex development disorders and patients with sexual deviations [2].

Sexism means gender discrimination, based on settings and beliefs claiming that differences between women and men establish the superiority of one sex over the other.
The society governed by men is called *patriarchal*, and the society governed by women is called *matriarchal*. At present, the entire world's known societies are patriarchal and, therefore, sexist. Persons supporting *sexim* ideology are called *sexists*. According to both sexist and racist ideologies, different populations are unequal as regards their physical and intellectual abilities; however, no convincing evidences showing the superiority on males over females or vice versa have been presented.

This term first appeared in the USA in the 1960s within the women's liberation movement called *female emancipation*. Especially frequently it was used in discussions of anti-female prejudice and *gender stereotypes*. *Feminism* and the struggle for female equality (female emancipation) conducted by the feminist movement are the main opponents of sexism. One of the manifestations of sexism is the society's greater intolerance of male homosexuality as compared to lesbianism. In countries where homosexual contacts are criminally liable, the corresponding law applies only to males. Similar situation was existed in Nazi Germany (Paragraph 175) and the USSR (Clause 121).

In its struggle with sexism, the European Council (2010) tends to use genderally neutral words and recommends that all 47 Member States should abstain from sexist terminology in their official documents. The appearance of this recommendation was caused by the Swiss deputy Doris Schtamp's statement requiring that female should not be described as "passive and second-rate creatures, mothers or sexual objects" (cited from the newspaper "Bild"). Switzerland was the first country to start implementing the changes: in the Swiss capital Bern, the words "mother" and "father" disappeared from official documents. Henceforward, the alternative word "parents" should be used.

Homosexualism (the Greek word "homos" means "similar") is sexual desire for people of the same gender.

The notion *homosexualism*, like the notion *homosexuality*, was introduced on a wide scale at the end of the XIXth century by Karl-Maria Kertbeny, an Austrian writer of Hungarian descent; it meant the sexual desire for people of the same gender and sexual relations with them.

Male homosexualism (gay), or *pederasty* (Ancient Greek: παις — "child", "boy", and ἐραστής — "loving"; literal meaning – "love of boys"). In Antiquity, pederastic relationships were

111

considered as superior of heterosexual the man-woman relationships, because only the pederastic relationships were thought to have intellectual and spiritual origins, while the man-woman relationships were viewed as purely carnal and lacking spiritual background. Of the Ancient Greek prosaic works addressing the topic of pederasty, the most famous work is the Plato's "Feast" stating the superiority of spiritual homosexual love of boys over the carnal man-woman relationships; this spiritual love was later termed "*the Platonic Love*".

At present, homosexual marriages are recognized officially and registered in the majority of civilized countries.

The American Psychological Association has provided scientific evidences confirming the expedience of same-sex marriages. According to Dr. Clinton Anderson, one of the leaders of this Association, values and benefits associated with homosexual marriage are similar to the values and benefits associated with traditional union.

In Nigeria and Zimbabwe, male homosexualism is prosecuted, and, in some countries, the punishment for homosexual acts is public death penalty.

Sapphism, Lesbian love or Lesbianism is female homosexualism. Homosexual women are called *lesbians*. This term originates from the word "Lesbos", the name of the Greek island where the Ancient Greek female poet Sappho was born and lived. Her poems were later regarded as a female homosexual love hymn. However, according to some ancient sources, Sappho also had sexual relations with men. Maximus of Tyre wrote that Sappho's relations with female students of her school were *platonic*. There are evidences for the existence of lesbian relationships in Ancient Sparta. In his description of Lakedaimon, Plutarch reported that "they attach so much importance to love that unmarried girls become erotic partners to women of noble families".

At the end of the XIXth century, female homosexuality remained practically unnoticed as compared to male homosexuality (the latter was prohibited by law and, therefore, was discussed in press). More information about the concept of female homosexuality became available after publication of the works of sexologists Karl Heinrich Ulrichs, Richard von Krafft-Ebing, Havelock Ellis, Edward Karpenter and Magnus Hirschfeld.

As soon as female homosexuality became a topic of discussions, it was described as a disease. Sigmund Freud in his work "Three articles on the theory of sexuality" (1905), called female homosexuality as "the inversion" and homosexual women "the inverts". He described the inverts as the women who have masculine characteristics. At that time, S. Freud used the notion of "the Third Gender" proposed by Magnus Hirschfeld (1899).

Magnus-Hirschfeld (14 May 1868 – 14 May 1935) was a German physician and sexologist. An outspoken advocate for sexual minorities, Hirschfeld founded the Scientific

Humanitarian Committee, which Dustin Goltz called "the first advocacy for homosexual and transgender rights".

After several years as a general practitioner in Magdeburg, in 1896 he issued a pamphlet Sappho and Socrates, on homosexual love (under the pseudonym Th. Ramien). In 1897, Hirschfeld founded the Scientific Humanitarian Committee with the publisher Max Spohr, the lawyer Eduard Oberg, and the writer Franz Joseph von Bülow. The group aimed to undertake research to defend the rights of homosexuals and to repeal Paragraph 175, the section of the German penal code that since 1871 had criminalized homosexuality. They argued that the law encouraged blackmail, and the motto of the Committee, "Justice through science", reflected Hirschfeld's belief that a better scientific understanding of homosexuality would eliminate hostility toward homosexuals.

Within the group, some of the members scorned Hirschfeld's analogy that homosexuals are like disabled people; they argued that society might tolerate or pity them, but never treat them as equals. They also disagreed with Hirschfeld's (and Ulrichs's) view that male homosexuals were by nature effeminate. It argued that male-male love is a simple aspect of virile manliness rather than a special condition.

The Scientific-Humanitarian Committee, under Hirschfeld's leadership, managed to gather over 5000 signatures from prominent Germans for a petition to overturn Paragraph 175. Signatories included Albert Einstein, Hermann Hesse, Käthe Kollwitz, Thomas Mann, Heinrich Mann, Rainer Maria Rilke,

August Bebel, Max Brod, Karl Kautsky, Stefan Zweig, Gerhart Hauptmann, Martin Buber, Richard von Krafft-Ebing and Eduard Bernstein.

The bill was brought before the Reichstag in 1898, but was only supported by a minority from the Social Democratic Party of Germany, prompting Hirschfeld to consider what would, in a later era, be described as "outing": forcing some of the prominent and secretly homosexual lawmakers who had remained silent out of the closet. The bill continued to come before parliament, and eventually began to make progress in the 1920s before the takeover of the Nazi Party obliterated any hopes for reform.

He saw himself as a campaigner and a scientist, investigating and cataloging many varieties of sexuality, not just homosexuality. He developed a system which categorised 64 possible types of sexual intermediary ranging from masculine heterosexual male to feminine homosexual male, including those he described under the word he coined "Transvestit" (transvestite), which covered people who today would include a variety of transgender and transsexual people.

In 1921 Hirschfeld organized the First Congress for Sexual Reform, which led to the formation of the World League for Sexual Reform.

Magnus-Hirschfeld was convinced that innovations in human sexuality can be transferred to the reforms in the society.

In 2011, official registration of the third gender was started in Australia. Since then, there have been three options for registration of gender in the Australian civil passport – namely, Male, Female and X-Gender (unidentified gender). The mark "X" registered in the passport can be used for transgender persons (transsexuals) and patients with congenital Disorders of Sexual Development (DSD), uncertain sexual orientation.

Transsexualism is a medical term denoting the condition of non-correspondence between the anatomical gender of a person and gender identity (i.e., psychic gender).

H. Benjamin (1953) had proposed the term *transsexualism* by who presented scientific description of this condition and defined it as "a personality pathology consisting in polar opposition of the biological gender/civil status and the psychic gender". This incongruous status causes a person to feel severe psychic discomfort called "gender dysphoria" accompanied by depression, which can lead to suicide [2].

Transsexualism is one of the psychic and behavioral Disorders of Self Identification (ICD-10 Code: F64.0).

Gender Identity Disorder can be diagnosed, if person's identification with the opposite gender is persistent (i.e., is present in the course of at least 2 years) and is not a manifestation of a psychotic disorder (for example, schizophrenia) or genetic anomalies (A.O. Bukhanovskiy, 1998).

The following types of transsexualism are distinguished:

➤ **Male-to-female transsexualism**, sex change from male to female,

➤ **Female-to-male transsexualism**, sex change from female to male.

Treatment method of Transsexualism is only Sex Reassignment Surgery (SRS) and hormonal correction in accordance with the patient's gender self-identification, including the change of documents and patient's socialization in new gender role.

Sex Reassignment Surgery is carried out after appropriate testing, diagnosis confirmation and exclusion of other disorders, including various endogenous diseases, schizophrenia, hysterical personality disorder, homosexuality, transvestism, alcoholism, narcomania, antisocial behavior, severe somatic disease and moderate or severe dementia. Also, the presence of registered marriage and children is taken into account in view of the involvement of corresponding judicial aspects [2].

Transsexualism is not directly related to sexual orientation: a transsexual person can be heterosexual, homosexual or bisexual in relation to his or her psychic gender. Since childhood, transsexuals struggle for harmony between their gender self-identification and their perception by social surroundings, for their *Egos* (selfishness) and for their right to change their gender. Inability to change their life situation and social adaptation problems may often provoke them to attempt suicide.

Bisexualism, or bisexual orientation, is one of the three possible sexual orientations. It is characterized by emotional, romantic (platonic), erotic (sensual) and/or sexual desire directed towards both same-sex and opposite-sex persons; these feelings are not necessarily equal in intensity or simultaneous. Bisexuality can be diagnosed in both males, females and transsexuals, and the respective persons are called *male bisexuals, female bisexuals and transgender bisexuals.*

Deviant manifestations of sexuality

Exhibitionism (the Latin word "*exhibeo*" means exposing, demonstrating) is a form of sexual behavior when sexual satisfaction is reached by demonstration of one's genital organs to unfamiliar persons (usually, to opposite-sex persons) and in public places. As a rule, an exhibitionist has no hopes for continuation of the contact or establishing a relationship.

Usually demonstration of genital organs is accompanied with sexual arousal, erection and publicly masturbation with orgasm and ejaculation. Appearance of exhibitionism as an individual phenomenon was contributed to by permanent wearing of clothes and its accompanying notion of nudity shamefulness.

Voyeurism (the French word *"voyeur"* means *spectator*), or "visionism" (the Latin word "visionis" means *view*), or "scoptophilia" (the Latin word *"scope"* means *review*) – is a desire to contemplate sexual intercourse or naked genital organs. For males, the view of naked female genital organs acts as a key stimulant that causes sexual arousal by an unconditional reflex, i.e., on the basis of a genetically determined program. Similar response to the view of genital organs is not characteristic of women; it is observed only in some women with previous favorable sexual experience and, thus, is an acquired reaction. For attaining pleasure from viewing the genital organs, one has to have sufficiently high sexual arousability.

Sexual fetishism (French: *fetichisme*) is a sexual behavior, in which various material objects (for example, clothes, footwear, various things made of a certain material, etc.) become a source and stimulus for sexual desire. As a rule, the objects that are related to opposite-sex persons become fetishized. For men usual fetishes include the women's panties, stockings, pantyhose, brassière and women's footwear. This type of fetishism is accompanied by fetishistic transvestism characterized by dressing up and wearing the clothing items mentioned above. In Japan, fetishism has become a source of income related to sex industry: unmarried girls practice selling the unwashed worn clothes, these clothes are packaged in air-tight bags and delivered to specialized shops where they are bought by fetishists. The parts of the body especially attractive for fetishists include hair, breasts, soles and legs. Specific features of genital organ morphology, details of appearance, human body anomalies or pregnancy can be fetishized. Photographs of the objects of interest can also play the role of fetishes.

Krafft-Ebing wrote: "Erotic fetishism is a condition whose psychologic motivation is characterized by fetishization of physical or even spiritual attributes of a person. Moreover, even the simple every day-use or similar objects can be fetishized and evoke strong associations with the person; also, they are always colored by vivid sensual sensations. Religious fetishism is similar to sexual fetishism in that religious fetishism also involves, depending on particular circumstances, fetishization of quite insignificant objects (e.g., nails, hair, etc.), which consecutively provoke the feelings that reach the level of ecstasy.

BDSM

BDSM: A composite acronym for "B&D" (bondage & discipline); "D&S" (dominance & submission); and "S&M" (sadomasochism). Used to refer to any consensual activities or lifestyles between adults which include some or all of these things. The term "BDSM" is used in a general sense to describe any situation or practice which includes erotic power exchange, dominance and submission, pain play, bondage, sensation play, or anything related to these.

BONDAGE: Any practice involving tying or securing a person, as with ropes, cuffs, chains, or other restraints. Restraint bondage, the most common form of bondage, involves immobilizing a person, by tying or otherwise restraining him or her to an object or by binding his or her limbs together. Stimulation bondage is any form of tying in such a way that the subject is not immobilized and has freedom of motion, but the ropes or ties shift and move against the body, often in sensitive or erogenous areas; certain forms of shibari are stimulation bondage. A person in bondage is said to be bound.

DISCIPLINE: 1. Any activity in which one person trains another person to act or behave in a specified way, often by enforcing rigid codes of conduct or by inflicting punishment for failure to behave in the prescribed way. 2. *Archaic* Any instrument used to enforce discipline or to punish physically, such as a whip or crop.

DOMINANT: A person who assumes a role of power or authority in a power exchange relationship. A dominant takes psychological control over or has power over another person, and may, for example, give that person orders which are to be obeyed. *Contrast* submissive; *see related* top, switch.

DOMINATE: To assume or exert control over; to take psychological power over. A person who controls another person or takes control of a scene is said to dominate that person. Dominant is a noun or an adjective; dominate is a verb. Domination, dominance: the act of wielding authority over another.

DOM: A dominant. *Usage:* Often indicates a male dominant; however, may be applied to a dominant of any sex.

TOP: One who administers some form of stimulation, such as spankings, floggings, or some other kind of stimulation on another person but does not have psychological control or power over that person. Contrast bottom; see related dominant.

SUBMISSIVE: One who assumes a role of submission in a power exchange relationship. A submissive is a person who seeks a position of or occupies a role of intentional, consensual powerlessness, allowing another person to take control over him or her. *Contrast* dominant; *see related* bottom, switch.

SUBSPACE: A specific state of mind that a submissive may enter, particularly after intense activities and/or (depending on the person) intense pain play, characterized by euphoria, bliss, a strong feeling of well-being, or even a state similar to intoxication. Thought to be related of release the endorphins in brain. The euphoria associated with subspace may last for hours or sometimes even days after the activity ceases.

SUBMIT: To give up power or control. A person who gives up power or psychological control to another is said to submit to that person. Submission: the act of giving up control.

BOTTOM: A person who receives spankings, floggings, or other forms of stimulation in situations which specifically exclude power exchange. For example, a masochist may be interested in receiving some kind of stimulation but may not be interested in giving up psychological control; whereas a submissive has given up authority and may receive some kind of stimulation on the instruction of a dominant, a bottom does not give up authority and may control exactly how, under what circumstances, and to what degree he or she receives some form of stimulation. *Contrast* top; *see related* submissive.

BOTTOM DROP: *Colloquial* A sudden, abrupt feeling of depression, unhappiness, or similar negative emotion in a submissive which may occasionally occur immediately after a period of BDSM activity. May include feelings of shame or guilt, especially if the submissive has traditional ideas about relationship or socially appropriate behavior; after a period of intense pain play, bottom drop may be related to the reduction of levels of endorphins in the brain as well.

S&M, SADOMASOCHISM: Any activity or practice involving the inflicting or receiving of pain; pain play.

SADIST: One who is aroused, excited, or receives sexual gratification from inflicting pain on another. *Contrast* of masochist. A sadist does not necessarily take pleasure in inflicting pain indiscriminately; for most sadists, the pleasure relies on knowing that the subject is also enjoying the experience.

MASOCHIST: One who experiences arousal, excitement, or sexual gratification from receiving pain. *Contrast* of sadist. Contrary to popular misconception, a masochist does not experience arousal at all forms of pain; stubbing a toe, for example, is unlikely to be arousing. The context of the pain is important.

MASTER: A dominant, usually in a TPE relationship. Usually male; the female equivalent is a mistress. *Contrast* slave.

SAFE, SANE, AND CONSENSUAL (SSC): A code of conduct which holds that any activity between adults is acceptable as long as it is safe, sane, and consensual. Often held up as a test to whether or not a particular activity is ethical. *See related* RACK. *Commentary:* Many people see a flaw in the idea of "safe, sane, and consensual" because whether or not an activity is "safe" and "sane" is subjective, and because people may choose to engage in activities which might not always be "safe," as in some forms of edge play (def. 1). (This is true even outside the BDSM community; consider skydiving, for example). Because of this, SSC has given way to the code of conduct called "RACK" (risk-aware consensual kink) in some places.

This glossary is a guide to many of the terms in the BDSM.

If you see something described in here which isn't to your tastes, it doesn't mean that you aren't or can't be interested in BDSM.

FRAME: A type of bondage furniture consisting of an upright triangle, usually made of wood and typically about seven to eight feet tall, sometimes with cross slats. A person can be bound to the frame with wrists together, arms above the head and affixed to the pinnacle of the triangle, and ankles bound to the base of the triangle with legs apart.

ABASIOPHILIA: *Psychology* Sexual attraction to people in or who use wheelchairs, casts, braces, or other orthopedic fixtures.

ABRASION: Any form of sensation play involving stroking or brushing the skin with rough, textured objects such as sandpaper, emery boards, and the like.

AFTERCARE: A period of time after intense BDSM activity in which the dominant partner cares for the submissive partner. *Commentary:* Some BDSM activities are physically challenging, psychologically intense, or both. After engaging in such activities, the submissive partner may need a safe psychological space to unwind and recover. Aftercare is the process of providing this safe space.

AGE PLAY: A form of role play in which a participant assumes the role of someone of a different age. ADULT BABY: *Colloquial* A person who engages in infantilism in the role of a very young child or infant. Typically, the submissive partner will assume the role of a very young (and hence powerless) child. *See related* adult baby, infantilism. *Commentary:* One common misperception about age play is that it appeals to pedophiles or is intended to simulate pedophilia. For those who engage in this activity, it is the powerlessness aspect of childhood and the inherent power imbalance between an adult and a child, rather than the childhood itself, that is appealing.

AGORAPHILIA: *Psychology* Sexual arousal from sex in public places. Does not necessarily involve exhibitionism; the excitement may come from the fear of being caught, rather than from being observed in a sex act.

ALGOPHILIA; *also,* ALGOLAGNIA: *Psychology* Sexual arousal from receiving pain.

ALPHA SUB: *Colloquial* In a relationship in which one dominant has more than one submissive, the submissive accorded the greatest power or respect among all the submissives. *See related* polyamory: "primary/secondary". *Commentary* Not all relationships which have more than one submissive include a hierarchy among the submissives; that is, not all such relationships have an alpha sub.

ANAL HOOK; *also,* BUTT HOOK: A smooth, blunt metal hook, typically about an inch thick and six inches long, with a small loop on one end. The shorter side of the hook is inserted into the anus, and a rope tied to the loop on the other end can be tied to an overhead fixture to force the wearer to kneel with his or her butt in the air, or can be secured to the wrists to bind the wearer's hands. Some anal hooks include a ball on the end that is inserted.

ANILINGUS: Any sexual activity involving licking, kissing, or tonguing another person's anus.

ANIMALISM: Any form of role play in which a participant assumes the role of an animal to be trained, such as a horse or dog. *See related* pony play, puppy play.

ANKLE CUFFS: Any cuffs specifically designed to be affixed to a person's ankles. Ankle cuffs are often made of leather, but may also be made of cloth, rope, metal, or even wood.

BANDAGE SCISSORS: Specialized scissors often used by emergency medical personnel, consisting of a pair of scissors with one sharp blade and one blunt blade with a rounded end. The blunt blade can be slid beneath bandages or anything else wrapped tightly around a limb without risk of cutting or injuring the person.

Commentary: Often used in BDSM to remove a person from tight bondage or mummification very quickly in the event of an emergency. A sturdy pair of bandage scissors will make quick work of even thick rope; a person totally wrapped in rope can be freed within seconds with bandage scissors without injury.

BASTINADO: Any form of pain play involving inflicting pain on the soles of the feet, often by striking, cropping, or whipping them.

BERKLEY HORSE: A type of bondage furniture consisting of a padded bench with integrated restraints and a pair of arms to which a person's legs can be affixed. A person is bent over the berkley horse and restrained in place; the berkley horse is designed so that it can be elevated, rotated, or moved into any position, and the arms to which the ankles are bound can be opened or closed. *Etymology:* The berkley horse was allegedly invented in 1828 by Theresa Berkley, a prodomme in London who specialized in flogging her clients.

BLINDFOLD: Any implement designed to prevent a person from seeing by covering the eyes. *Also, verb* the act of using a blindfold on a person.

BODY BAG: A long, heavy bag, often shaped like a narrow sleeping bag and typically made of canvas, rubber, or latex, used to restrain a person very tightly. Sometimes includes integrated straps which wrap around the person within the bag. *See related* mummification.

BODY HARNESS: A harness consisting of a series of straps designed to be worn around the torso, which may optionally include a mechanism for locking the harness into place and may also include rings or other attachments for ropes, cuffs, or chastity belts.

BODY MODIFICATION: Any practice, including piercing, tattooing, branding, and the like, intended to modify, often permanently, the appearance of one's body.

BOI: *Colloquial* 1. A person, usually biologically female and often boyish or "butch" in manner, appearance, or dress, who is submissive; commonly but not exclusively used in lesbian. 2. An effeminate man.

BONDAGE BELT: A belt used to restrain a person, which consists of a heavy band of leather or a similar material which can be strapped or locked about the waist and which has several attachment points to which the subject's wrists may be bound.

BREAST BONDAGE: A specific form of bondage involving binding around or over the breasts. *Also* breast press, karada, shinju.

BUTT PLUG: A sex toy intended for anal stimulation, consisting of a flared dildo, usually quite short, with a wide base, designed to remain securely in the anus until removed.

CARABINER: Any device used to connect two chains or ropes together, often in the form of a D-shaped metal ring with a spring-loaded lever which can open the ring. *See related* panic snap. *Commentary:* Carabiners are not usually appropriate for suspension, as they cannot easily be removed if the suspended person's full weight is bearing down on them.

CATHERINE'S WHEEL: A large, upright wheel, usually made of wood, to which a person may be bound and then rotated to any position.

CHASTITY: The practice of disallowing any form of sexual release or sexual activity, sometimes imposed on a submissive by a dominant. Some forms of imposed chastity include the use of locking devices such as chastity belts to prevent direct sexual stimulation of the genitals. Sometimes called chastity play, enforced chastity.

CHASTITY BELT: Any device intended to prohibit contact with or stimulation of the genitals. Female chastity belts often take the form of a lockable harness which passes between the legs and around the waist; male chastity belts may include a locking enclosure into which the penis is placed.

CHASTITY PIERCING: Any body piercing intended to prevent sexual intercourse; as, piercing along the labia which can be locked together to prevent penetration, or a piercing of the foreskin which can be used to pull the foreskin over the head of the penis and lock it in place.

CLOVER CLAMP: A specific type of nipple clamp consisting of a clamp with a lever mechanism to which a chain or cord is affixed in such a way that pulling on the chain or cord increases pressure on the clamp.

COLLARING CEREMONY: A formal ceremony celebrating or symbolizing a commitment between a dominant and a submissive, typically during which a collar is placed around the submissive's neck.

CONSENT: Affirmative permission, assent, or approval. In a BDSM context, "consent" is an affirmative assent to engage in a particular activity, freely given without coercion or distress. Informed consent: Consent freely given with full and prior knowledge of the conditions and potential consequences of the assent. *Also, verb* To give affirmative permission to engage in an activity.

CONSENSUAL NON-CONSENT: Any situation in which one person knowingly and voluntarily gives up the ability to prevent another person from doing whatever he or she wants; as, for example, deliberately engaging in activities which the submissive may be physically prevented from resisting and does not have a safe word. Some forms of rape play are consensual non-consent. *Commentary:* Consensual non-consent is still consent. A person who gives consent in this way is giving affirmative assent to engage in an activity that he or she will not be able to stop in the middle; it can be thought of as consenting to an activity in such a way that the consent may not be revoked.

CONTRACT: A mutually negotiated, written agreement between a dominant and a submissive, outlining the submissive's limits, the activities the participants wish to explore, and the like. *Commentary:* BDSM contracts are not legally valid or enforceable, but are useful tools for defining what activities are and are not acceptable and in what contexts.

CONTRAPOLAR STIMULATION: *Physiology* Of or relating to any form of stimulation which produces both pleasure and pain sensations simultaneously.

CORPORAL PUNISHMENT: Any activity involving disciplining a person through physical means, as by spanking.

CROSS-DRESSING: Sexual arousal or gratification from wearing clothing appropriate for the opposite sex.

CUCKOLD: One whose partner practices cuckoldry. *Usage:* Most often refers to a man whose partner practices cuckoldry by having sex with other partners, though occasionally used to describe women as well. For the female equivalent, *see* cuckquean.

CUCKOLDRY: The practice by which a dominant takes one or more sexual partners other than his or her submissive, for the purpose of humiliating the submissive. *Commentary:* Cuckoldry is distinct from the practice of »polyamory« in the sense that it is done in a context where the submissive has no direct control over the dominant's other partners, and the primary purpose is to humiliate the submissive. Those who are aroused by cuckoldry are most often attracted to the humiliation and powerlessness aspects of it. The majority of the people who practice cuckoldry as a sexual fetish are women, who humiliate their male partners by having sex with other men.

CUCKQUEAN: One whose partner practices cuckoldry. *Usage:* Always refers to a woman whose partner has sex with other people.

CUFF: 1. Any restraint which has a band or band-like structure, which may be made of metal or of a flexible material such as canvas or leather, intended to be strapped or locked around an extremity such as a wrist or ankle for the purpose of securing or immobilizing it. 2. *Archaic* the fist. *Also, verb* 1. To restrain or immobilize by means of a cuff or cuffs. 2. To strike a rapid blow, as with the hand. 3. *Archaic: Cuff with*, to engage in a fistfight with.

DACRYPHILIA; *also,* DACRYLAGNIA: *Psychology* Sexual arousal from seeing a partner cry.

DEVICE BONDAGE: *Colloquial* Bondage involving highly specialized equipment, furniture, or devices, often very elaborate, to immobilize a person.

DIAGONAL CROSS: *See* St. Andrew's cross. *Usage:* Primarily British; uncommon in the United States.

DOOR HANGAR: A n implement for bondage consisting of a short piece of flat webbing, usually an inch or two wide, connected to a thick metal or wooden dowel on one end and a metal ring on the other. The webbing can be hung over the top of a door; when the door is closed, the dowel prevents the webbing from being pulled through the door, and a person may be bound to the ring.

DRAGON'S TAIL: An unusual type of whip consisting of a handle, often made of wood and wrapped with leather, to which a wide triangular piece of thin leather or suede is attached. This leather or suede forms a lash which is a hollow tube tapering to a point at the striking end.

DRAGON'S TONGUE: An unusual type of whip consisting of a handle, often made of wood and wrapped with leather, and a lash made of a single wide piece of leather or suede wrapped around another, thinner suede lash. The outer lash is rolled into a tube around the inner lash, and tapers to a point at the striking end.

ENDORPHINS: Naturally-occurring opiate-like chemicals produced in the brain in response to pain, which block pain and can produce a euphoric sensation. The euphoria sometimes described by people who engage in BDSM is often attributed to endorphins.

ENGLISH: *Archaic* caning.

EXHIBITIONIST: One who is sexually aroused by showing others his or her body or by being watched, particularly in a sexual setting or while engaged in sexual activity.

EXHIBITIONISM: The act of engaging in exhibitionistic behavior, such as sexual behavior, for the sexual gratification of the person being watched.

FACESITTING; *also,* FACE SITTING: *See* queening.

FEAR PLAY: Any BDSM activity centered around creating the feeling of fear in one or more of the participants. This does not necessarily have to involve elements of pain or actual danger; sensations of fear may be created through the use of psychological pressure, sensory deprivation, the threat of pain or discomfort, or even exposing a person to the object of a phobia.

FEMINIZATION: The practice of enforcing activities or behaviors on a male submissive which are typically associated with women, as cross-dressing, requiring the submissive to sit when urinating, and the like. Often used as a form of humiliation play. Also referred to as sissification.

FETISH: 1. Formally, *Psychology* a non-sexual object whose presence is required for sexual arousal or climax; informally, anything not generally considered sexual which arouses a person, as a *foot fetish* or a *leather fetish.* 2. Anything of or relating to BDSM in general; as a *fetish convention, a fetish event.* 3. Items, practices, or apparel relating to BDSM; as, *fetish photography, fetish clothing.*

FIGGING: The practice of placing a piece of carved ginger root into the anus or vagina. The result is a burning sensation which many people claim can intensify orgasm, and which other people use as an adjunct to physical discipline such as spanking. *Commentary:* This practice is believed to date back to Victorian times, when it was used in conjunction with caning as a technique for disciplining errant women.

FIRST GIRL: In the Gor novels, one of a group of female sex slaves owned by the same master, who is considered 'first' or predominant over the other sex slaves. In Gorean D/s, a woman who identifies as a slave and has status or rank over any other women who consider themselves slaves to the same person.

FISTING: The practice of inserting the entire hand into the vagina or (less commonly) into the anus. *Commentary:* Vaginal fisting is actually quite a bit easier to do than most people realize; the human body is quite accommodating. Contrary to common misconception, fisting is not done by making a fist and shoving it into the vagina; rather, the fingers are placed together and inserted slowly; as the hand is inserted, the fingers tend to curl into a loose ball. Many women experience intense orgasms from vaginal fisting.

FISTING SLING: A sling designed in such a way that the person within the sling is reclined with the legs spread apart, in a posture convenient for fisting.

FLAGELLATION: A generic term for any sort of activity involving flogging or whipping.

FLAGGING: The act of wearing a specific clothing, insignia, jewelry, or other sign as a means of expressing interest in a specific form of BDSM activities. *See related* hanky code.

FLOGGER: An implement used to strike a person, consisting of a handle with multiple lashes attached to it. The lashes are typically made of leather, but may also be made of materials such as rope, suede, horsehair, or even Koosh balls. *See also* cat, cat o' nine tails, dread koosh flogger, fire whip.

FLORENTINE: A flashy flogging technique involving the use of a flogger in each hand. The floggers are swung in a figure-8 pattern. Sometimes called a double weave. *See related* triple weave.

FORCED ORGASM: An orgasm induced in a person against that person's will or as part of resistance play, often by means of bondage combined with sexual stimulation. *See related* consensual non-consent.

FOUNTAIN OF VENUS: *Colloquial* Water sports involving urination by a woman.

FREEPLAY: BDSM activities in which there is no domination or submission. *See related* top, bottom, sensation play.

FROG TIE: A specific form of bondage in which the person kneels and the ankles are bound to the thighs, preventing the person from rising; the wrists are then bound to the ankles.

FUCKING MACHINE: Any device or machine which is designed to simulate the act of sex; often consisting of a dildo affixed to a reciprocating motor so as to thrust in and out of a person. Many varieties of fucking machines exist, some designed so that the subject straddles or sits on them, others designed to be used when the subject is prone or spread-eagle.

FUNNEL GAG: A gag, usually consisting of an oblong or penis-shaped rubber or plastic bit, which has a tube running through it connected to a funnel. When the gag is placed in the mouth, any liquids introduced into the tube will pass into the mouth, and the person wearing the gag has no choice but to swallow them.

GAG: Any device or object designed to be placed in the mouth, most commonly to prevent a person from speaking or making loud sounds, sometimes to hold the mouth open. *Also, verb* 1. To place an object into the mouth to prevent a person from speaking. 2. To choke, particularly by placing something in the mouth.

GOLDEN SHOWER: A form of water sports involving the act of urinating on a person.

GOR: A mythical planet created by science fiction writer John Norman and used as the setting for an entire series of science fiction novels. The novels describe a civilization in which women occupy an extremely submissive position in society and are often used as sex slaves. The novels describes a formalized, ritualized set of social structures centered around female submission and male superiority, which have been adopted by a subcommunity of people within the BDSM community.

GOREAN D/S: Male domination and female submission according to a formal system adapted from the fictitious society described in the Gor novels, and characterized by strong hierarchy, male superiority, and an elaborate system of protocols. Includes such elements as ritualized postures and positions which women are expected to take in the presence of men under certain circumstances. Also Gorean master, Gorean slave: one who adopts a dominant or submissive role in a manner which reflects the society described in the novels. *See related* kajira, first girl.

GREEK: *Colloquial* Of or related to anal sex.

GROPE BOX: A long, narrow, enclosed box, often made of wood, with many openings along its front and sides, into which a person may be placed and then groped or fondled by people outside the box. A person within a grope box is helpless to prevent the fondling and often cannot see the people doing the fondling.

GWENDOLINE HOOD; *also,* SWEET GWENDOLINE HOOD: A specific type of hood, often made of leather or latex but sometimes made of materials such as PVC or Spandex, which fits around the head and covers the mouth but has a large opening for the eyes and nose. *Etymology:* Named for Sweet Gwendolyn, a character in a series of bondage comics written by artist John Willie in the 1950s and 1960s, who was often depicted wearing this style of hood.

GYNARCHY: 1. Formally, *Sociology* A political or governmental system ruled by women. 2. *Colloquial* Femdom, particularly femdom in which all females are assumed to be superior to the male.

HARD LIMIT: A limit which is considered to be absolute, inflexible, and non-negotiable. *Contrast* soft limit.

HOBBLE SKIRT: A item of clothing consisting of a very tight skirt that ends below the knee, which prevents freedom of motion of the legs, allowing the wearer to walk slowly in a hobbling motion but not to move quickly.

HONOR BONDAGE: *Colloquial; see* psychological bondage.

HOOD: Any covering designed to go over the head, often partly or completely covering the face as well.

HORSE: 1. A piece of bondage furniture consisting of a plank supported by two legs on each end, similar to a sawhorse. A person may be bent or tied over the horse and flogged or spanked. 2. *See* wooden horse.

HUMILIATION PLAY: Sexual arousal from activities which include an element of humiliation, shame, or embarrassment for one or more of the participants. *Commentary:* Humiliation play is a relatively unusual taste that is often very difficult to explain to someone who doesn't understand it. While humiliation play may carry little or no risk of injury, it can be psychologically very intense, and is sometimes the psychological equivalent of edge play.

IMPALEMENT: A practice in which a person is bound, usually while standing, and penetrated anally or vaginally with a dildo attached to the end of a fixed pole or rod in such a way that the person cannot escape or remove himself or herself from the dildo. *Commentary:* This practice can be dangerous if not done correctly. The person must be bound in such a way that he or she cannot fall if he or she loses balance.

INFANTILISM: A type of role play in which one of the adult participants takes on the role of an infant, and may be dressed in diapers, suck on a pacifier, and so forth. *See also* age play; *see related* adult baby.

INFIBULATION: 1. Chastity piercing, particularly of men. 2. In some cultures, the practice of female genital mutilation, typically forced on women at the onset of puberty and often for religious reasons. The practice consists of surgical removal of the clitoris and/or sewing the labial lips together to prevent sexual penetration. 3. In some historical contexts, particularly in ancient Rome, the practice of sewing the foreskin over the head of the penis to prevent a male slave from engaging in sexual intercourse.

KENNEL PLAY: A specific form of puppy play in which the submissive is confined to a kennel or doghouse as part of the play. *See related* animalism.

LIFESTYLE: 1. *Colloquial; often "the lifestyle"* Of or pertaining to involvement in BDSM, as in *How long have you been in the lifestyle?* 2. Of or pertaining to a TPE relationship, as in *We practice lifestyle D/s*.

LIMIT: A boundary, which may be set by a dominant or a submissive, which specifies a point past which any activity will not go. *See* soft limit, hard limit. *See related* edge play.

MASTIGOTHYMA: *Psychology* Sexual arousal from being flogged.

MERINTHOPHILIA: *Psychology* Sexual arousal from being tied up. *See related* bondage bunny.

METAFETISHIST: *Colloquial* 1. A person who is aroused by introducing a partner to new sexual activities. 2. A person with a wide range of different sexual turn-ons or tastes. 3. A person who enjoys learning or exploring new sexual activities.

MESSALINA SYNDROME; *also,* MESSALINA COMPLEX: A seldom-used synonym for nymphomania. *Etymology:* Derives from the alleged promiscuous sexual appetite of Valeria Messalina, the wife of the Roman Emperor Claudius.

MILITARY PLAY: A specific form of role play which involves military-style settings, uniforms, hierarchy, or protocol.

MILKING: 1. The practice of stimulating the male prostate, often with a finger or with an implement such as a dildo, or of stimulating the perineum in such a way as to produce ejaculation without orgasm. 2. The practice of inducing orgasm repeatedly in a man, often by sexually stimulating him over and over, until he is no longer able to produce ejaculate. 3. Stimulating the prostate by means of an electrode built into a dildo or similar probe, inserted into the anus and connected to an electrical stimulation device such as a TENS unit. The electrode causes involuntary contraction of the muscles around the prostate, causing ejaculation without arousal or orgasm.

MISTRESS: Female equivalent of a master.

MUMMIFICATION: A form of bondage in which the subject is immobilized by being entirely wrapped quite tightly, as with Saran wrap, rope, fabric, or similar material.

NEWBIE: *Colloquial* A newcomer to BDSM; or, more generally, a newcomer to any sport, hobby, or subculture.

NIPPLE CLAMP: Any clamp or clamp-like device designed to be clamped to a subject's nipples. May include a mechanism for adjusting or limiting the amount of pressure applied to the nipple. Clothespins make good (and cheap!) nipple clamps. *See related* clover clamps, tweezer clamps.

ODALISQUE: (Literally, Turkish *oda* chamber, room + *liq* woman) *Archaic* A female sex slave.

ORGASM CONTROL; *also,* ORGASM DENIAL: The practice whereby one person is not permitted to reach sexual orgasm without the permission of another person, or for a set period of time, or sometimes at all, even though that person may be permitted (or required) to engage in sexual activity or sex.

ORIENTATION PLAY: Any activity in which a person is ordered or instructed to engage in sexual activity with another person whose sex is not appropriate for the first person's sexual orientation or identity, as for example instructing a straight female to engage in sexual activity with another woman.

PADDLE: Any stiff, hard implement, often made of wood, used for striking a person, most commonly on the buttocks. *Also, verb* to strike with a paddle.

PAIN PLAY: Any activity in which one person inflicts pain on a consenting partner, for the pleasure of one or more of the people involved. Spanking, flogging, paddling, whipping, and so on are all forms of pain play. *See related* sadist, masochist.

PAINGASM: *Colloquial* An orgasm achieved through painful stimulation. *See related* contrapolar stimulation.

PANIC SNAP: A specific type of carabineer designed in such a way that the mechanism can be opened to release a rope or chain even if a full weight is bearing down on it.

PANSEXUAL: 1. Of or relating to all sexual orientations, sexes, and gender identities. 2. One who engages in sexual or erotic activities with partners of all sexes and orientations. Pansexual event: an event catering to people of any sexual orientation or identity. Pansexual group: any group open to membership by any person regardless of sex, sexual orientation, or sexual identity.

PARACHUTE: A small leather cone with a hole in its center, which is often used in CBT. The parachute is wrapped around the scrotum, and weights are suspended from it, pulling on the scrotum and compressing the testicles.

PERCUSSION: Any form of impact play involving striking with a blunt or fairly heavy implement, such as a buck hammer or Taylor hammer.

PERVERTIBLE: *Colloquial* Any object which serves an ordinary and prosaic function, but which also has a use in BDSM activities. For example, clothespins are often used as nipple clamps; saran wrap can be used for mummification; paint stirrers are sometimes used as paddles; and so on. *Etymology:* This term was coined by David Stein.

PETTICOATING: *Archaic* feminization, especially, feminization done by making the subject dress in women's clothing as a punishment for some transgression.

PILLORY: *Archaic; see* stock. *Also, verb, archaic:* to expose to scorn or ridicule.

PILLORY BED: A bed with a stock built into the headboard and/or footboard, such that a person lying on the bed may have the stock closed over his or her ankles, wrists, or wrists and head to restrain the person and prevent him or her from leaving the bed.

PONY PLAY: An activity in which the submissive takes on the role of a pony; for example, by walking on all fours, sometimes with a bit or bridle in the mouth; by pulling carts; by allowing the dominant to ride on his or her back; and so on. *See related* animalism, mouth bit.

POSTURE COLLAR: A specific type of high, rigid collar, often shaped to the wearer's neck, which prevents the wearer from moving his or her neck and forces the wearer to hold his or her head high. *See related* corset collar.

POWER EXCHANGE: Any situation where two or more people consensually and voluntarily agree to a power relationship in which one (or more) people assume authority and one (ore more) people yield authority. This relationship may be for a predetermined time, or indefinite. Relationships based on indefinite power exchange are often referred to as TPE relationships. The defining factor of power exchange is the conscious, deliberate construction of a power dynamic in which at least one person assumes psychological control to some agreed-upon extent over at least one other person.

PROSTATE MILKING: *See* milking (def. 1).

PROTOCOL: Any defined, enforced code of behavior which a submissive is expected to abide by. A protocol often imposes constraints and limits on the submissive's behavior, particularly in social settings; for example, a protocol may specify that a submissive is not to speak to another person without the dominant's permission, may not speak unless spoken to, and so on.

PSYCHOLAGNY: *Psychology* Orgasm without physical stimulation. *Commentary:* In some D/s relationships, the dominant partner will train the submissive partner to orgasm on command, with a word or a gesture, but without being touched physically.

PSYCHOLOGICAL BONDAGE: Bondage without the use of ropes or other restraints, in which the submissive is simply commanded not to move.

PUNISHMENT TIE: Any form of bondage done in such a way that the bound person's pose or the bondage itself is painful or uncomfortable, or any kind of bondage done with the intent of causing pain or discomfort to the bound person. Some forms of shibari include punishment ties. *See related* pain play.

PUPPY PLAY: An activity in which the submissive takes on the role of a puppy, as by barking, walking on all fours, and in some cases even sleeping in a doghouse or cage. *See related* animalism, kennel play.

QUEENING: A practice whereby a dominant, usually but not always female, sits on the face of a submissive, who is often restrained, forcing the submissive to perform oral sex and/or anilingus. Sometimes may also include breath control. *See related* queening stool.

RAPE PLAY, *also* RAPE FANTASY: A form of role play in which one person stages a mock "rape" for the purpose of gratification of all the people involved. *See related* consensual non-consent, resistance play. *Commentary:* A surprisingly common form of BDSM play, often staged so as to fulfill a woman's sexual fantasies of rape or coerced sex in a safe and controlled way.

RESISTANCE PLAY: Any mutually consensual activity in which one person struggles against another and is subdued by "force." May involve rape play; some forms of bondage include resistance as well. *See related* consensual non-consent.

RHABDOPHILIA: *Psychology* Sexual arousal from being whipped, beaten, or flogged.

RIMMING: *Colloquial* Anilingus.

SAFE CALL: A practice sometimes used as a safety measure when meeting a new partner for the first time. The safe call is a prearranged telephone call made to a trusted friend at a specific time to let that friend know that everything is okay; may involve the use of special code words to indicate whether or not the person making the safe call is in danger or distress.

SAFEWORD: A predefined "code word" which a submissive can use to stop an ongoing activity if it becomes too much. *Commentary:* Safewords are often used in situations such as

141

resistance play, where the submissive may be expected to struggle or resist and where the word "no" might not actually mean no. In such cases, for safety's sake it's often helpful to have some word that *does* mean "no," and is a word unlikely to come up otherwise.

SCARIFICATION: A form of body modification involving cutting the skin, often in intricate or elaborate patterns, in such a way that the healing process leaves behind a permanent scar.

SCAT; *also* SCAT PLAY: Any activity involving feces. *Commentary:* Very likely to elicit a squeak reaction from most people.

SCENE: 1. A specific period of BDSM activity; as in, *We had a scene lasting about two hours last night.* 2. *Colloquial* The BDSM community as a whole. 3. *In the scene:* participating in the organized BDSM community.

SELF-BONDAGE: The act or practice of tying one's self up or otherwise restraining one's self, sometimes as a part of masturbation. Often includes some mechanism by which the person may be freed after a set amount of time, which may include a timer mechanism to release a key or otherwise release the person.

SENSATION PLAY: Any BDSM activity involving creating unusual sensations on a person, who may be blindfolded, as with ice cubes, soft fur or cloth, coarse materials, and the like. Sensation play is much more mild than pain play and may or may not include an element of power exchange.

SENSORY DEPRIVATION: Any practice intended to reduce a person's ability to see, hear, or use his or her other senses, either to create a psychological state of arousal or fear or as part of sensation play. *See related* ball hood, blindfold, isolation hood.

SESSION: *See* scene (def. 1). *Usage:* Most often used to indicate a scene with a prodomme.

SHIBARI: A type of bondage originating in Japan and characterized by extremely elaborate and intricate patterns of rope, often used both to restrain the subject and to stimulate the subject by binding or compressing the breasts and/or genitals. Shibari is an art form; the aesthetics of the bound person and the bondage itself are considered very important. Also sometimes called kinbaku.

SHOE FETISHISM: A type of sexual expression centered around a fixation on shoes, sometimes as part of submission. Submissive shoe fetishism may involve acts such as licking, kissing, or caring for the dominant partner's shoes.

SISSY: A male submissive subject to feminization, as by being made to wear women's clothing, act like a woman in social settings, and so on.

SISSIFICATION: *also* feminization.

SLAVE: A submissive, usually in a TPE relationship. Contrast master. *Commentary:* People who self-identify as "master" or "slave" often see dominance or submission as a cornerstone of their identity, an essential part of who they are as people; this self-identify may affect and inform almost every aspect of their lives.

SLING: Aim item of furniture, usually made of leather, canvas, or nylon webbing, suspended by chains or cables from the ceiling. A person may sit in the sling and arranged for easy availability to such activities as sexual intercourse, fisting, and the like. Slings may include additional mechanisms to restrain the person within the sling or to keep the legs spread apart. *See also* fisting sling.

SOFT LIMIT: A limit which is not necessarily be set in stone, but which may be flexible or may change over time. *Contrast* hard limit. *See related* edge play (def. 2). *Commentary:* One of the most powerful aspects of BDSM is that it offers a way for people to challenge their soft limits, testing themselves against their own boundaries in a safe and controlled way.

SOMNOPHILIA: *Psychology* A Sexual arousal from having sex with a sleeping person.

SOUND: A thin, solid metal rod designed to be inserted in the urethra, often as a part of a medical role play.

SPECULUM: A medical instrument commonly consisting of two or occasionally three probes designed to be inserted into the vagina or (less commonly) the anus, together with a mechanism intended to spread the probes apart, opening the vagina or anus. Sometimes used in medical role play scenarios.

STING: A sensation of quick, sharp pain. *Usage:* The feeling caused by being struck by a flogger is usually described in terms of thud or sting.

STRAP: An implement used for striking a person, consisting of a long, flat piece of heavy leather. *Also, verb* to strike with a strap.

STRAP-IN: A dildo designed to penetrate a person either vaginally or anally and then be held in place by a strap or harness, sometimes equipped with a lock to prevent it from being removed.

STRAP-ON: A dildo attached to straps, a harness, or some other mechanism designed to be worn around the waist.

STRAPPADO BONDAGE: A specific bondage technique in which a person's hands are tied behind his or her back, then a rope is tied to the wrists and attached to an overhead fixture or pulley tightly enough so that the bound person is forced to bend over with his or her arms in the air. *Commentary:* This is a physically demanding form of bondage which exerts strain on the arms and shoulders and may be dangerous if done by people who are not experienced and knowledgeable.

SUSPENSION: Any form of bondage in which the person bound is suspended partially or completely off the floor, often by ropes affixed to an overhead point (as with a kotori in shibari), or by means of a rigid bar with attached suspension cuffs.

SWITCH: 1. One who can change roles, being either dominant or submissive (or, less frequently, sadistic or masochistic) at different times or with different partners. 2. A thin, flexible rod, often made from a green branch of a tree such as a willow tree, used for striking people; similar to a cane. 3. *See* polyamory: switch«. *Also, verb* 1. To change roles, as from a dominant role to a submissive role. 2. *(infrequent)* To strike with a switch (def. 2).

For the religious rites of ancient Greece, see Orgia. For the American synth rock band, see Orgy (band).

In modern usage, an orgy is a sex party where guests freely engage in open and unrestrained sexual activity or group sex.

SWINGER: Swingers' parties do not always conform to this designation because at many swinger parties the sexual partners may all know each other or at least have some commonality among economic class, educational attainment or other shared attributes. Some swingers contend that an orgy, as opposed to a sex party, requires some anonymity of sexual partners in complete sexual abandon. Other kinds of "sex party" may fare less well with this labelling.

Participation in an "orgy" is a common sexual fantasy and group sex targeting such consumers is a subgenre in pornographic films.

SYBIAN: One popular variety of commercially-available fucking machine consisting of a dildo affixed to a dome-shaped saddle which the user sits on. *Commentary:* The Sybian has been described by a friend of mine as "a machine that rips orgasms out of women." After my experiences watching people use these machines, I have to agree.

TOP DROP: *Colloquial* A sudden, abrupt feeling of depression, unhappiness, or similar negative emotion in a dominant which may occasionally occur immediately after a period of BDSM activity. May include feelings of guilt, especially if the dominant believes he or she has made an error, or has traditional ideas about relationship or socially appropriate behavior.

TPE, TOTAL POWER EXCHANGE (TPE): A relationship in which one person surrenders control to another person for an indefinite duration, and in which the relationship is defined by the fact that one person is always dominant and the other is always submissive.

One of the more extreme forms of power exchange. Sometimes referred to as lifestyle D/s. *See related* master, slave.

TRANSVESTITE: One who engages in cross-dressing.

TWO-COLUMN TIE: In bondage, any form of tie which binds two parts of the body together; as, for example, any tie which ties the wrists together, or ties a wrist to an ankle. *See related* one-column tie.

VAMPIRE GLOVES: Gloves used for sensation play which have a large number of short spikes or needles protruding from the palms and/or fingers.

VANILLA: *Colloquial* Not interested in or involved with BDSM or activities related to BDSM; as, *a vanilla person. Usage:* Sometimes considered condescending or insulting.

VINCILAGNIA: *Psychology* Sexual arousal from tying up or otherwise physically restraining a partner.

VIPER: An instrument used for striking a person, consisting of a rigid handle and a small number of narrow, flat lashes made of thin rubber, each of which tapers to a point at the striking end.

VOYEUR: One who is excited or aroused by watching others, particularly in a sexual context or while engaged in sexual activity.

VOYEURISM: The act of engaging in voyeuristic behavior. *See related* polyamory: candaulism.

ZENTAI SUIT: A skin-tight full-body suit, often made of Spandex or a similar breathable material, that includes a full hood, most often without openings for the eyes or mouth. Such suits are often worn as part of a sexual fetish.

Any person may use some sexual practice and devices according appetence.

- Eye bandaging with a semi-transparent cloth.

- Phalloimitators of different lengths (ranging from 6.0 cm to 30.0 cm) and thicknesses (ranging from 2.0 to 6.0 cm in diameter) for vaginal sex shaped like the penis, with penile head. These goods are made mainly of latex, plastic, metalloceramics or wood and are characterized by various degrees of elasticity (i.e., they are elastic, hard or rigid); also, they are covered by various coating materials (i.e., they are smooth or supplied with numerous knobs, ribs, horns, etc.); the can vibrate.
- Phalloimitators (the Greek word "phallos" means "penis") for anal sex (smaller in diameter); they can be with or without a penile head.
- Anal toy beads.
- Phallos-shaped vibrators for anal and vaginal sex.
- Rubber dolls (females or males) for sexual intercourse. The sex doll models that vibrate, produce voluptuous moans and sounds during frictions and have containers for sperm collection have been created. Fashion designers have created intimate clothes for sex dolls.

"Not all the recommendations that can be found in the medical science deserve introduction into practice.

The science covers the universe, while application of methods covers but individual cases".

Vattsyayana Mallanaga [1].

147

I am really interested in legitimate right for all forms of sexual behavior, which I did, and which tends to be determined of natural impulses.

marquis Donatien Alfons François de Sade, 1772

Salvador Dali. Engraving for novel **"Justine ou les Malheurs de la vertu"**, marquis de Sade, 1782

You should be strongly sensitive to be felt.

Nikkollo Paganini, 1820

7. Erogenous zones

Erogenous zones (from Greek *Eros* - love and *genous* - producing) are different area of the human body with heightened sensitivity, the stimulation of which may result in the production of enjoyable and delectation senses, sexual arousal and orgasm.

Specific erogenous zones are associated with sexual response, and include areas of the genitals, notably the foreskin and corona of the glans clitoris and rest of the vulva, peryanal skin and lips.[1] The rete ridges of the epithelium are well-formed and more of the nerves are close to the external surface of the skin than in normal-haired skin.[1]

Nonspecific sensitive zones are similar to normal-haired skin and have the normal high density of nerves and hair follicles. These areas include the mouth and lips, sides and back of the neck, the inner arms, the sides of the thorax, inguinal area, feet et al., and may include full body surface as well.

Sexual function and sexual desire are subjects with quite large individual variation.

Diverse factors of the environment, health status, family relationships, ethnic factors, culture, aesthetic perception, intellectual level and many other factors play a role in the appearance of sexual desire and erogenous receptiveness.

Sexual desire is closely related to psychic status of the individual. In a broad sense, sexual life encloses all the somatic and psychic processes that are related to reproduction biology. In a narrow sense, sexual life implies coitus with its accompanying set of psychosomatic reactions.

Distant receptors and sensory organs cause significant effects on sexual function due to their stimulation by the environment [10]. Their important role in appearance of sexual arousal was emphasized by Sigmund Freud. Thus, for example, the organs of vision contribute to arousal by telling the brain cortex about the beauty of the sexual object. Delights of the sexual object's body cause pleasure and provoke/augment sexual arousal.

Stimulation of erogenous zone`s tactile receptors of a person in a non-aroused state (for example, breast skin of a woman) may cause the feeling of pleasure and, simultaneously, the appearance of sexual arousal demanding the augmentation of pleasure. Appearance of enjoying demands the extension of gratification as one mechanisms of arousal [21, 22].

The role played of erogenous zones in these processes is well known. All of them act to bring definite pleasure when appropriately stimulated; the pleasure increases the tonus providing the motor energy necessary for sexual intercourse completion [21, 22].

The tactile stimuli cause very strong effects. Reaching the brain, various tactile stimuli produce sexual feelings and vivid fantasies. Via the higher nervous activity, various stimuli jointly cause the body to prepare for sexual reaction. Sexual tonus manifests itself through each element of male and female behavior, including gait, face mimics, mood, etc.. [2, 5, 11, 13]. A significant role is played by the activity of sex glands. According to Russian gynaecologist A. Mandelshtam, "...there is certain functional parallelism between the gonads and the sexual resonance". [19].

A very important role is also played by acoustic stimuli (e.g., music, voice) and visual stimuli (pictures, hair style, manner of dressing, underwear, woman's appearance and woman's attractiveness). Also, significant role is played by smell function.

Russian academician Ivan P. Pavlov wrote that smell plays a determining role in sexual arousal in many animals (I.P. Pavlov, Complete works. Moscow-Leningrad, Edition of the Academy of Sciences of the USSR, 1951, v. IV, pp. 25-26) [10]. Use of perfumes containing attractive aromas augments the feelings, associated with sex partner physical closeness and emphasizes his or her individuality and character. Pleasant healthy body odor is a strong erogenous stimulus. Smell centers are part of the limbic system whose activity ensures human emotions, perception, subconscious reactions and mood.

G.S. Vasilchenko provided the following illustration of the leading role played by the brain cortex in the sexual arousal: "... in conditions of negative mental attitude, penile (or clitoral) head rubbing against clothing causes neither erection nor positive subjective feelings, whereas in conditions of romantic setting even pain stimuli can be perceived as pleasant." [5].

The direction of libido is an inborn characteristic, but it changes and forms in the process of accumulation of personal experiences.

Sexual arousal appears not only due to the effects of hormones or reception of distant stimuli, but also due to tactile stimulation of especially sensitive zones of the body, *erogenous zones.*

Erogenous zones are the very sensitive zones of skin or mucosal membrane (for example, genital organs, breasts, lips, oral cavity) whose stimulation causes sexual arousal or orgasm. Also, these zones can include neck, internal surfaces of thighs, back, abdomen (especially in the region of the navel) and other regions.

The erogenous zones are divided, according to their importance, into the *main* and the *accessory* zones. All the main erogenous zones are located similarly in the majority of people, while the accessory zones are usually subject to a greater inter-individual variation.

In these zones, according to some authors (Dietz K., Hesse P. G., 1974; Obgartel L., 1974), specific receptors of genital sensitivity the so-called "genital bodies" (Latin: "corpuscula nervosa terminalis genitalia") that respond to pressure are located in addition to the ordinary tactile, temperature and pain receptors. Possibly, location of these genital bodies determines the location of the erogenous zones. In males, the most sexually excitable zone is the penile head; scrotum is excitable to a lesser degree. Also, sexual arousal is caused by penile skin displacement or application of pressure to the penile root.

In females, the main erogenous zones include: breasts, especially nipples, internal surfaces of the thighs, pubic region, perineum, pudendal lips, clitoris, G-spot and anus. Many other zones of the body can also become erogenous zones.

Clitoris is a small midline erectile body that is homologous to the male penis (see Chapter 3). It consists of a glans, body, and crura. The glans is easily visible at the superior junction of the labia minora, and a covered fold of skin as *capuche*. The body, consisting of paired corpora with an incomplete septum between them, is 1 to 2 cm wide and 2 to 4 cm long [1]. The crura, 5 to 9 cm in length, are attached to the ischial bone and covered by the ischiocavernosus muscle [1].

During arousal, its cavernous bodies are filled with blood, which results in erection, the clitoral response to the excitatory stimulus.

Clitoral and penile smooth muscle appears to share a similar neuroregulatory mechanism. Erotic sensitivity of the clitoris depends on individual variation: some women develop an erotic response to application of rhythmic pressure to the whole clitoral region, while other women respond only to application of slow strokes to the clitoral body or quick strong and rhythmic pressure to the body or head of the clitoris, with pulling the clitoral head upward or pressing it against the pubic bone. Some women require frequent change of the zone of stimulation: several tens of seconds after the

155

beginning of stimulation, the zone of the clitoris being stimulated becomes in excitable, while the adjacent zone of the clitoris acquires high excitability. In many women, clitoral stimulation evokes erotic reaction only after preliminary general caresses increasing the genital excitability.

Clitoris, especially the *glans clitoris* and the surrounding zone, is richly supplied with nerve endings – approximately 8,000 per unit 1 cm; this is one of the zones of the woman's body most sensitive to stimulation. In this zone, number of the sensory endings per unit area is 3-4 times as much than in the *glans of penis*.

In this view, clitoris is very responsive even to changes in temperature or very delicate stimulation. Clitoris is very responsive of weasels, even of contact through clothes. Many women casually obtain extremely pleasant sensations while riding a horse or a bicycle or in a trainer hall. Clitoral caresses without penetration can result in an orgasm.

As a rule, the term "clitoris" is used to denote only the small visible part of the clitoris located in the upper third of the space between the large pudendal lips. However, the greater part of the clitoris, like the underwater part of an iceberg, is hidden deep within the large pudendal lips that form the borders of the vaginal entrance. The body of the clitoris divides into two crura that pass within the base of the respective pudendal lips; numerous nerve endings are located there. Statistical data shows that more than 50 % of women consider clitoris as the main erogenous zone.

The most sensitive part of the clitoris is the clitoral glans; however, its stimulation can result in too strong, nearly painful, sensations whose intensity depends on the degree of arousal and, accordingly, the clitoral erection.

Clitoral *erection* (i.e., clitoral *enlargement* and *straightening*), like penile erection, results mostly from the cavernous bodies filling with blood. While filling with blood, the clitoral cavernous bodies *erect* (i.e., enlarge and become more rigid). However, the cavernous bodies, though directly related to clitoral glans erection, have mostly the autonomic sensitivity [5, 8]. During erection, the clitoral glans that is also abundantly supplied with arterial vessels increases in diameter and becomes filled with arterial blood; clitoral glans surface smoothes out, and location of sensory nerve endings changes to be more superficial; skin becomes thin; and clitoral erogenous sensitivity significantly increases. Intensive stimulation of the clitoral glans before clitoral erection causes pleasant sensations; however, sensitivity of the erected clitoris becomes so high that even a slight and gentle tactile stimulation results in strong erogenous impulses that pierce like an electric current. Sometimes, these reactions become very intensive, resembling the pain sensations. For male partner, it is very important to detect this change in sensations and, accordingly, make a timely shift to another method of stimulation or vaginal intercourse without additional clitoral stimulation. Skillful weasels and diverse manipulations with the clitoris can lead to an unforgettable orgasm and ecstasy even before the vaginal penetration.

Awkward or careless manipulations at the peak of arousal can lead to negative consequences, including the interruption of the "arousal pearl string". In this view, in case of any doubts, the responsibility for the "qualified care" as regards the clitoral weasels may be delegated to the woman herself or the woman's female partner. Possibly, this is one of the substantiated arguments in support of the *lesbian couples*. For a male sex partner, it is very important to preserve and increase his phallic power, because the woman acquainted with the real delight associated with vaginal penetration will never consent to limit her sexual activity to clitoral stimulation, whatever pleasure is brought by it.

According to its associated pleasure sensations, the clitoral orgasm resulting from clitoral stimulation, oral sex or women's masturbation is often regarded as more powerful and acute, but shorter and more superficial, as compared to vaginal orgasm.

Clitoris may be corrected by plastic surgery. In case of decreased clitoral sensitivity caused by folded skin cover of the clitoris, a functional operation can be used to expose the clitoris. Also, aesthetic operations can be used to enlarge or decrease the clitoris. The clitoris contains numerous nerve endings and blood vessels; thus, for successful outcome of the surgical clitoral plasty, high professional level of the surgeon and careful conduct of the operation are necessary.

Sometimes, clitoral piercing decoration is used; this procedure not only causes the cosmetic and distracting effects, but augments the sensations associated with vaginal and oral sex. However, according to the opposite viewpoint, genital piercing decreases the sex organ sensitivity due to skin scarring. The puncture aimed at passing a micro rod or ring is made either in the skin fold above the clitoris (i.e., the clitoral hood) or in the center of the clitoris or nearer to its base. Presence of the clitoral piercing necessitates taking account of the possibly higher sensitivity and the necessity to comply with appropriate hygienic requirements.

Small **pudendal lips** form the lateral borders of the vaginal vestibule (vulva). Mucous membrane of the internal surfaces of the small pudendal lips is very richly supplied with the nerve endings sensitive to erogenous stimulation (see the description of small pudendal lip anatomy in Chapter 3)

Anatomically, small pudendal lips have direct skin connections with the clitoral glans and clitoral fold; however, the small pudendal lips are considerably inferior to the clitoris as regards the erogenous sensitivity and do not show the same high local erogenous selectivity as found in the clitoris.

Erogenous sensitivity of the small pudendal lips is ensured by sensory reception on their internal surfaces and the deep visceral sensitivity of the clitoral cavernous bodies. The clitoral crura composed of cavernous bodies with their accompanying vasculonervous bundles pass within the base of the respective pudendal lips and border the pudendal cleft and, accordingly, the vaginal opening on both sides together with *musculus bulbocavernosus* and *musculus ischiocavernosus* (see Chapter 3).

Clitoral erection involves the erection of its cavernous bodies located within the basis of the respective pudendal lips; this results in some narrowing of the vaginal entrance and sharp increase in the pudendal lip erogenous sensitivity.

After the erection of the clitoris and its crura within the basis of the respective pudendal lips, each penile penetration of the vagina causes intensive pleasure sensations and woman's desire to be penetrated deep into the vagina, with variable intensity and force of frictions.

As regards the achievement of orgasm, many women note that more frequent gliding of the penis, with *catching* on the cavernous bodies of the vaginal vestibule, causes clitoral erection and "orgastic cuff" formation. Long period of elongated frictions results in production of impulses causing sustained pleasure and leading to the peak of passion and orgasm. Alternating use of deep/prolonged and shallow/frequent frictions is especially effective. Intensive "processing" of this zone causes expressed inflow of blood to the vagina and increase in the vaginal mucosa sensitivity.

The **vagina** is a muscular sheath that connects the uterus and the external genitalia. It has both sexual and reproductive functions and this dual function is reflected in various aspects of its structure, especially its vascular supply, its sensory and motor innervation and its marked distensibility. The wall of the vagina consists of an inner lining of stratified squamous epithelium, an intermediate layer of smooth muscle and an outer adventitial layer. The lining of the vagina is folded into numerous rugae which are related to its extreme distensability. The vaginal entrance is closed by the bulbocavernosus muscle, a striated muscle.

Histological studies of the vagina have shown that the vaginal epithelium is intimately related to a dense capillary network in the subepithelial layer [10]. This vascular network may be related to the production of transsudate during sexual arousal, caused *lubrication*. Somewhat deeper, the loose submucosal layer contains a complex of numerous large veins and smooth muscle fibers that resemble cavernous tissue. The smooth muscle layer consists of an inner circular layer and a strong external longitudinal layer. The adventitial layer consists of connective tissue and a large plexus of blood vessels.

Female sexual arousal is a neurovascular phenomenon involving nerve-regulated vascular reactions. Neurovascular interactions during sexual arousal lead to several hemodynamic phases which affect simultaneously the clitoris, vestibular bulbs, and labia minora as well as the vagina. In the resting phase, the vagina is a sheath containing a potential space with a minimal blood flow and very low oxygen tension in the wall [17,18]. The earliest detectable sign of sexual arousal based on studies with experimental models is a significant increase in vaginal wall and clitoral blood flow [19]. There is a significant increase in clitoral cavernous pressure as well [19]. With the onset of increased vaginal blood flow, production of vaginal transsudate ensues [10,20]. A significant rise in tissue oxygen tension follows about 20 seconds later which indicates increased inflow of arterial blood. In humans, vaginal and labial oxygen tension increases from 4 to 8 times baseline during sexual stimulation [17,18] and vaginal wall blood flow increases approximately threefold [21-23]. In humans, clitoral blood flow has been estimated to increase from 4 to 11 times baseline during sexual stimulation [24]. The increased blood flow reaches a plateau phase during which vaginal fluid transsudate production continues. The final or resolution phase is characterized by the slow return of blood flow to baseline values. In women, up to 20-30 minutes is required for vaginal oxygen tension to return to baseline [17].

Posterior **vaginal fornix,** anterior **vaginal fornix** – so called **Gräfenberg-spot** (G-spot) and the **uterine cervix** become the locus of vaginal sensitivity: applying pressure to these zones can lead to multiple undulating orgasmic contractions of the uterus. After appropriate stimulation and reaching the peak of pleasure, prolonged lance-like thrusts and pressure in this zone result in the deep orgasm, the *orgasmic status* (Latin: *status orgasmicus*). It is necessary to be cautious and attentively monitor the woman's reactions, because too strong and excessive thrusts can cause painful, unpleasant sensations and lead to vaginism and couple disharmony. Sometimes, these painful sensations can be erroneously recognized as the manifestations of passion (especially if occurring at the peak of pleasure). Detailed description of the signs of pleasure experienced by women is presented below (Chapter 8).

The vagina forms part of maternal passages; it is nearly deprived of the nerve endings similar to those ensuring small pudendal lip and clitoral somatic innervation and, thus, erogeneity. However, this does not mean that the vagina is devoid of sensitivity. The vaginal sensitivity is another type of sensitivity, called visceral sensitivity that belongs to autonomous nervous system. Mucous, submucous and muscular coats contain numerous sensitive neuron plexuses of autonomic (i.e., sympathetic and parasympathetic) nervous system. Nerve plexuses are located mostly in the region of uterine cervix and vaginal fornixes. Receptors of autonomic neurons respond to rhythmic pressure applied to this region, as well as to changes in tissue perfusion and distension. Visceral sensory signals are transmitted to autonomic nervous ganglia and then to central nervous system where an autonomic response in the form of changes in the heart rate and blood pressure, dilation of blood vessels in the pelvic and other regions, as well as secretion of neurotransmitters and endorphins, the pleasure hormone, is initiated. Vaginal erotic sensitivity is subject to a significant inter-individual variation.

Nevertheless, many women note high vaginal responsiveness to exciting stimulation. As a rule, vaginal stimulation during masturbation or sexual intercourse directly and indirectly causes clitoral reaction (for example, via the anatomical zone located immediately below the clitoris or due to pulling the small pudendal lips) resulting in sensations inside or near the entrance to the vagina; in view of this, the vaginal erogenous zone is very sensitive to stimulation. This is just the rhythmic pressure (frictions) during vaginal penetration of the penis against the uterine cervix that causes the beginning of orgasmic sensations in more than 30 % of women.

In 1950, Ernst Gräfenberg, a German gynecologist, described the region in anterior part of vagina, as responsible zone for self-stimulating females. This zone was later called "**Gräfenberg's spot**", or "**G-spot**", was popularized in 1980 by sexologist Beverly Whipple as a tribute to the memory of gynecologist *Ernst Gräfenberg* [175]. Some females recognized the anterior vaginal fornix adjacent to the uterine cervix is supersensitive part of vaginal wall for initiating orgasm in consequence of masturbation and deep vaginal coitus.

Female breasts, especially the **nipples** and the pigmented area around the nipple called *areola*, also belong to the main erogenous zones. There is a strong direct reflectory relationship between the degree of nipple zone excitation and the woman's general sexual arousal. Also, during the arousal phase, there is some kind of nipple "erection" resulting from increased smooth muscle tone in this zone. Nipple stimulation causes *oxytocin* secretion by posterior pituitary lobe of *hypophysis*; by acting on smooth muscles of the uterus and the vagina, *oxytocin* increases the myometrial tone and influences on the uterine contractions during an orgasm. Orgasm intensity is directly proportional of oxytocin secretion.

Locations of *additional erogenous* zones are subject to a significant individual variation; they include face, lips, neck, abdominal and dorsal skin surface, feet, medial surfaces of thighs, mons pubis, perineum, anus, etc.. Erogenous zone sensitivity can change depending on the method of stimulation, e.g., its fondling with hand, lips, tongue, genital organs, body surface or different objects. Also, erotic sensations depend on the circumstances, mood, well-being and degree of sexual arousal. In the course of sexual intercourse and the arousal phase, some erogenous zones can become more sensitive, while other zones lose their sensitivity.

Stimulation of erogenous zones during sexual intercourse or masturbation results in increasing sexual arousal, erection and orgasm/ejaculation. In many cases, combined stimulations of these zones significantly increases the "sharpness" of sensations during an orgasm.

The list of female erogenous zones and the discussion of flirt variants are infinite!

To learn them, it is not sufficient to merely get acquainted with anatomical data: it is also necessary to have sufficient imagination power, aesthetic taste and wish.

The whole female body is able to act as an erogenous zone!

Erogenous zone assessment is, perhaps, the most pleasant and interesting of those activities that require a vivid and natural response! ☺

Try a different way and start experimenting.

Erogenous zones are not the smoothly operating command buttons. It all depends on the nuances.

Sometimes, a sparkle of desire is enough to produce a fire.

Knowledge of self-setting erogenous zones and pleasurable sensations is important for each female.

Many sexologists think it appropriate to recommend regular self-stimulation of clitoris, nipples, small pudendal lips and vaginal vestibule to their female patients. This corresponds to the so-called "delighting concept" that forms the basis of the sex therapy proposed by Professor Helen Kaplan Singer [26, 44].

To increase the sensitivity of erogenous zones, it is important to use not only a delighting approach, but preliminary relaxation of both the body and mind as well.

The main erogenous centers in a woman are her brain, mind and feelings!

Decreasing the acuity of senses is quite unnecessary, because it leads to anesthesia. The most interesting experiences are conscious!

Vattsyayana teaches that:

"Those inflamed with sexual desire require variety to support it.

Thus, the lovers equipped with the right skills stimulate each other's desire.

Sexual intercourse necessitate the multiplicity of approaches at least similar to that needed in bow shooting or any other weapon handling skill".

"Ever action must be followed a response;

a stroke should be produce responsive beat,

a kiss should be induce a sympathetic kiss". [1].

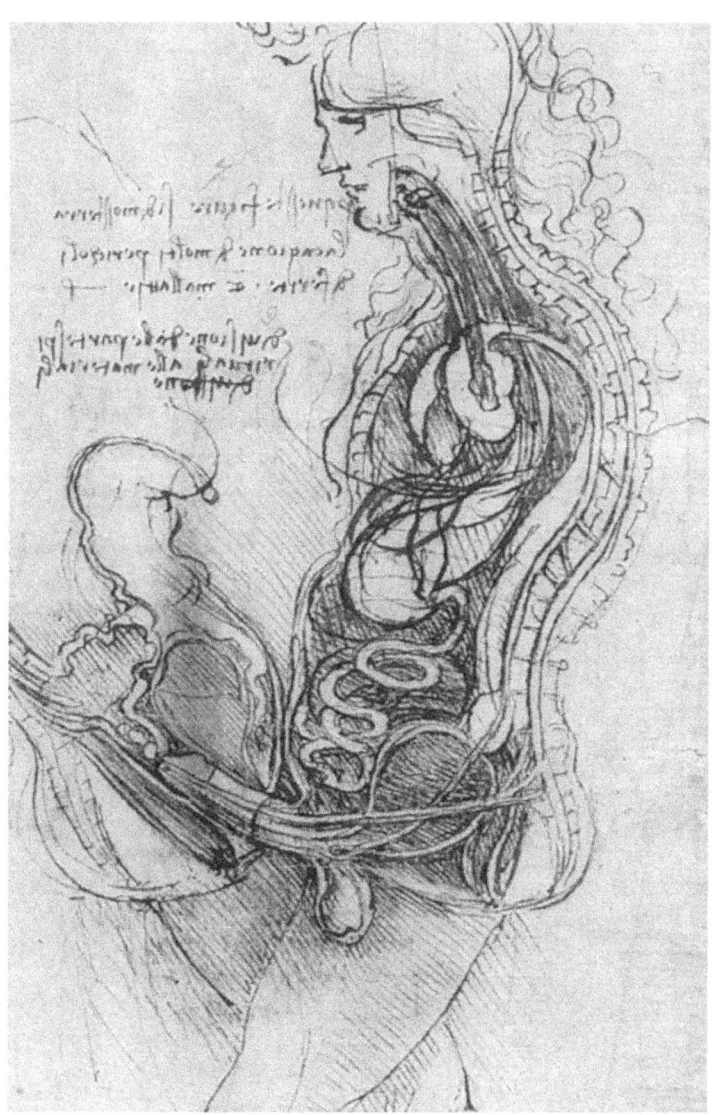

"I expose men to the origin of their first, and perhaps second, reason for existing."

Leonardo Da Vinci, 1493 **" Copulation"**

Sex is a form of love in an era where is no time for love.

Sigmund Graff, 1969

8. Penetration and sex senses

Penetration or sexual intercourse (coitus, copulation) in heterosexual pairs, commonly denotes the insertion and thrusting of a male's penis into a female's vagina. Sexual intercourse is the only physiological process that performs with 2 partners.

Coitus is derived from the Latin word coitio or coire, meaning "a coming together or joining together" or "to go together" and is usually defined as penile-vaginal penetration.

Penetration by the hardened, erect penis is additionally known as intromission, or by the Latin name emission penis (Latin for "insertion of the penis"). Copulation, although usually used to describe the mating process of non-human animals, is defined as "the transfer of the sperm from male to female" or "the act of sexual procreation between a man and a woman".

Other sexual penetrative acts may performed: anal sex, oral sex, fingering, or use of a strap-on dildo. Non-penetrative sexual practice has been referred to as "outercourse", but may also be among the sexual acts contributing to human bonding and considered intercourse.

"Vanilla sex" (or conventional sex) is a description of what a culture regards as standard or conventional sexual behavior. Different cultures, subcultures and individuals have different ideas about what constitutes this type of sex.

There is a variety of possible ways and positions for performing a sexual intercourse.

Anal and oral sex may be regarded as sexual intercourse, but they, as well as non-penetrative sex acts, may also be regarded as maintaining "technical virginity" or as "outercourse," regardless of any penetrative aspects. Heterosexual couples often engage in these practices not only for sexual pleasure, but as a way of avoiding pregnancy and maintaining that they are virgins because they have not yet engaged in penile-vaginal sex.

Sexual intercourse between non-human animals is more often referred to as copulation; for most non-human animals, mating and copulation occurs at the point of estrus (the most fertile period of time in the female's reproductive cycle), which increases the chances of successful impregnation. However, bonobos, dolphins, and chimpanzees are known to engage in sexual intercourse even when the female is not in estrus, and to engage in sex acts with same-sex partners. Like humans engaging in sex primarily for pleasure, this behavior in the aforementioned animals is also presumed to be for pleasure, and a contributing factor to strengthening their social bonds.

Likewise, some gay men view frotting or oral sex as maintaining their virginity, with anal penetration regarded as virginity loss, while other gay men consider frotting or oral sex to be their main forms of intercourse. Frot is a slang term derived from frottage (ult. from the French verb frotter, "to rub") describing a form of non-penetrative male/male sex that usually but not always involves direct penis-to-penis contact.

The term was originally popularized by gay male activists who disparaged the practice of anal sex, but has since evolved to encompass a variety of preferences for the act, which may or may not imply particular attitudes towards other sexual activities.

Lesbians may regard oral sex or fingering as loss of virginity, and may also regard tribadism as a primary form of sexual activity. Tribadism (TRIB-ə-diz-əm) or tribbing, commonly known by its scissoring position, is a form of non-penetrative sex in which a woman rubs her vulva against her partner's body for sexual stimulation, especially for ample stimulation of the clitoris. This may involve female-to-female genital contact or a female rubbing her vulva against her partner's thigh, belly, buttocks, arm, or other body part (excluding the mouth). A variety of sex positions are recorded, including the missionary position.

Female-female genital sex is not exclusive to humans. Females of the bonobo species, found in the Democratic Republic of the Congo, also engage in this act, usually known as GG rubbing (genito-genital). "Perhaps the bonobo's most typical sexual pattern, undocumented in any other primate, is genito-genital rubbing (or GG rubbing) between adult females. One female facing another clings with arms and legs to a partner that, standing on both hands and feet, lifts her off the ground." In bonobos, the clitoris is larger and more externalized than in most mammals. Bonobos rub their clitorises together rapidly for ten to twenty seconds, and this behavior, "which may be repeated in rapid succession, is usually accompanied by grinding, shrieking, and clitoral engorgement". On average, female bonobos engage in genital-genital rubbing "about once every two hours" [180-182].

Anal sex involves stimulation of the anus, anal cavity, sphincter valve or rectum, mostly commonly employing the insertion of a man's penis into another person's rectum. Oral sex consists of all the sexual activities that involve the use of the tongue, rest of the mouth and throat to stimulate genitalia. It is sometimes performed to the exclusion of all other forms of sexual activity, and may include the ingestion or absorption of semen or vaginal fluids. Fingering is the manual (genital) manipulation of the clitoris, vulva, vagina, or anus for the purpose of sexual arousal and sexual stimulation. It may constitute the entire sexual encounter or it may be part of mutual masturbation, foreplay or other sexual activities.

Sexual intercourse often ends when the man has ejaculated, and thus the partner might not have time to reach orgasm. In addition, premature ejaculation (PE) is common, and women often require a substantially longer duration of stimulation with a sexual partner than men do before reaching an orgasm.

Coitus may be classified:

Fast coitus and *prolonged penetration* depending on its duration

Coitus reservatus - saved intercourse (Latin *reservatus* - reserved, saved) - a form of sexual intercourse in which a man tries to restrain orgasm for prolong the erectile stage and keep himself in higher phase of arousal as long as possible avoiding the seminal emission (without ejaculation).

"Immissio Membri Virile In Vaginae Sine Ejaculatio Seminis" (Introduce the penis in the vagina without ejaculating the semen). In this practice, orgasm is separated from ejaculation, making possible enjoyment of the pleasure of sexual intercourse without experiencing seminal ejaculation, while still experiencing orgasm.

The psychologist Havelock Ellis writes: "Coitus Reservatus, – in which intercourse is maintained even for very long periods, during which a woman may have orgasm several times while the penetrative partner succeeds in holding back orgasm, – so far from being injurious to a woman, is probably the form of coitus which gives her the maximum gratification and relief".

Coitus interruptus, also known as the rejected sexual intercourse, withdrawal or pull-out method, is a method of birth-control in which a man, during intercourse withdraws his penis from a woman's vagina prior to orgasm (and ejaculation). The man then directs his ejaculate (semen) away from his partner's vagina in an effort to avoid insemination. Perhaps the oldest documentation of the use of the withdrawal method to avoid pregnancy is the story of Onan in the Torah. This text is believed to have been written down over 2,500 years ago. It has been suggested that the pre-ejaculate ("Cowper's fluid") emitted by the penis prior to ejaculation normally contains spermatozoa (sperm cells), which would compromise the effectiveness of the method. However, several small studies have failed to find any viable sperm in the fluid. While no large conclusive studies have been done, it is now believed the primary cause of method (correct-use) failure is the pre-ejaculate fluid picking up sperm from a previous ejaculation. For this reason, it is recommended that the male partner urinate between ejaculations, to clear the urethra of sperm, and wash any ejaculate from objects that might come near the woman's vulva (e.g. hands and penis).

Sexual intercourse often ends when the man has ejaculated, and thus the partner might not have time to reach orgasm. In addition, *premature ejaculation* (PE) is common, and women often require a substantially longer duration of stimulation with a sexual partner than men do before reaching an orgasm.

Masters and Johnson found that men took about 4 minutes to reach orgasm with their partners; women took about 10–20 minutes to reach orgasm with their partners, but 4 minutes to reach orgasm when they masturbated.

Scholars Weiten, Dunn and Hammer have reasoned, "Unfortunately, many couples are locked into the idea that orgasms should be achieved only through intercourse [penetrative vaginal sex]. Even the word foreplay suggests that any other form of sexual stimulation is merely preparation for the 'main event.'... ...Because women reach orgasm through intercourse less consistently than men, they are more likely than men to have faked an orgasm [185, 186].

According to the National Survey of Sexual Health and Behavior (NSSHB), in 2010, men whose most recent sexual encounter was with a relationship partner reported greater arousal, greater pleasure, fewer problems with erectile function, orgasm, and less pain during the event than men whose last sexual encounter was with a non-relationship partner.

With regard to adolescent sexuality, sexual intercourse is often for relational and recreational purposes as well. However, teenage pregnancy is usually disparaged, and research suggests that the earlier onset of puberty for children puts pressure on children and teenagers to act like adults before they are emotionally or cognitively ready, and thus are at risk to suffer from emotional distress as a result of their sexual activities. Some studies have concluded that engaging in sex leaves adolescents, and especially girls, with higher levels of stress and depression. A majority of adolescents in the United States have been provided with some information regarding sexuality, though there have been efforts among social conservatives in the United States government to limit sex education in public schools to abstinence-only sex education curricula.

Czhechian sexologists interviewed of 1000 women in order to identify the factors influencing on the vaginal orgasm frequency.

The following questions were asked:
What is the preferred duration of preliminary fondling and the coitus itself?
What is your usual degree of concentration on your vaginal sensations during the penile-vaginal interaction?
Does the probability of an orgasm increase with increasing penile size?
It was found that 21.9 % of the interviewed women had never had a vaginal orgasm.
This interview showed that vaginal orgasm frequency increases with prolongation of sexual intercourse, female's concentration on her vaginal sensations, partner's penile size, sexual education level and experience.
Preliminary foreplay causes practically no effect on the frequency and intensity of vaginal orgasm [31].

Erica Jong as a third wave feminist published in USA a novel "Fear of Flying" in 1973 [154], became a bestseller and classic work for a whole generation of American housewives and enriched the English language with the expression "Zipless Fuck" that means a spontaneous and anonymous sex without emotional involvement, love or attachment.

According of Erica Jong, "Reconciling our growing feminism with unacceptable lust for male body is a big problem. This combination is not easy to endure. On the one hand, it becomes clear that men are actually afraid of women, either more openly or latently. This is a torment to be a liberated female's desire to having sex, and masculine with soft penis. All the greatest historical events fade when compared to the quintessential combination of an ever eager woman and a permanently soft penis." [154].

Alfred Kinsey [23, 24] was the first investigator to carry out an appropriate statistical analysis and state that even the couples considered as happy have very poor or restricted sex.

In 1991, scholars June M. Reinisch and Ruth Beasley of the Kinsey Institute stated, "The truth is that the time between penetration and ejaculation varies not only from man to man, but from one time to the next for the same man." They added that the appropriate length for intercourse is the length of time it takes for both partners to be mutually satisfied, emphasizing that Kinsey "found that 75 percent of men ejaculated within two minutes of penetration. But he didn't ask if the men or their partners considered two minutes mutually satisfying" and "more recent research reports slightly longer times for intercourse".[62] A 2008 survey of Canadian and American sex therapists stated that the average time for intromission was 7 minutes and that 1 to 2 minutes was too short, 3 to 7 minutes was adequate and 7 to 13 minutes desirable, while 10 to 30 minutes was too long [183-185].

William Masters, much earlier than the Kinsey's studies, noted that, for some married couples, sexual intercourse is as rare as casual event. This sex regimen is unable to satisfy a woman and, thus, should be replaced with a more optimal variant.

At least 2000 years ago, ancient Taoist healers wrote books that frankly and clearly described love and sex. Taoists were not lustful or shy, because they viewed sex as a necessary element of physical and mental health and well-being in males and females [9].

Taoist physicians viewed sex as a part of the natural order of things. Sex was not only enjoyed and relished, but also considered as wholesome and life-extending. Taoist philosophy advocates unification with one's own infinite natural power through patience, relaxation and achievement of naturalness. The "Tao of Love" was developed on the basis of this natural philosophy of foresight, natural sexual energy preservation and flexibility.

Today, the concept of the "Tao of Love" seems unpracticable; however, with each new discovery of Western sexologists and scientists, its prescriptions become increasingly acceptable.

The main Taoist principles: <u>sexual intercourse prolongation by male partner using ejaculation control, as well as exercising the art of satisfying a woman – have become the important aspects of the methods used for the achievement of sexual harmony</u>!

The conceptual idea of prolonged *delectation* of woman for satisfaction required a literally sexual revolution for the West to accept it. However, in Ancient China about 2500 years ago, this idea formed the basis of Taoist philosophy. Probably, the appearance of the "Tao of Love" made the Chinese society recognize the change from *matriarchate* to *patriarchate*. Consecutive Chinese history has been characterized by presence of remarkable cases of longevity, and almost all of these cases resulted from practicing the "Tao of Love".

The *five features* of a woman's desire and arousal described in the "Tao of Love" [9] are as follows:

o   Her face blushes and her ears become hot.

It means that her mind has been conquered by thoughts about making love. At this moment, he should start the sexual intercourse at a moderate pace and in a teasing manner; he should enter her with a shallow movement and wait for her response.

o   Her nose becomes covered in sweat, and her nipples swell.

It means that her passion's fire has somewhat increased. Now, the nephrite stem can be pushed to the full depth of the trough, but not deeper than its length. He should thrust deeper only after an additional increase her passion.

- Lowered tone of her voice and appearance of dry and hoarse sounds coming from her throat are the manifestations of increased passion. Her eyes are closed, and her tongue protrudes from her mouth; she breathes frequently and more noticeably.

  At this stage, his *"nephrite stem"* (penis) may loosely pass in and out. The sexual intercourse gradually reaches the ecstatic phase.

- Her *"red sphere"* (vulva) becomes copiously lubricated, and the passion's fire approaches its peak; every thrust produces the *Yin flow*.

  His nephrite stem slightly touches the *"water chestnut reef valley"* (deep vagina). Now, he may use the following sequence: one thrust directed to the left, one thrust directed to the right, one slow thrust and one quick thrust. Alternatively, he can use some other method at his discretion.

- When her *"gold lotus"* (legs) flex as if embracing him, her fire and passion reach the peak level. She wraps her legs around his waist and holds his shoulders and back with her hands; her tongue remains protruded. Upon appearance of these signs, he can pass his penis deep in the deep chamber (vagina). These deep thrusts allow her to reach whole-body ecstatic state.

Taoism contains recommendations related not only to development of friction styles, but the proper number of *frictions* for ideal sexual intercourse as well. According to ancient literature, real satisfaction of a woman can be reached only after one thousand penile *frictions*.

Perhaps, this figure, if assessed too formally by a man unfamiliar with the Tao of Love, can be regarded as a requirement whose fulfillment transforms a pleasurable activity into a hard work. However, the persons who experienced in the "Tao of Love" practices consider this requirement as a source of inspiration rather than a hard work. The *Idea* of "1000 Steps" (friction) can be easily accomplished in the course of 30 minutes (i.e., 1800 seconds) at a quite moderate speed. Knowing one's own ability to provide such an outstanding degree of satisfaction to his sex partner is an especially pleasant feeling in itself. This knowledge of own ability to satisfy even the most demanding woman increases self-esteem in a man.

According to Western sexologists, the idea of "One Thousand Steps" (i.e., frictions) is implementable. Ability to have sex that lasts 6 to 10 minutes is considered a reasonable criterion of male potency. In the course of this period, a man with normal potency is able to carry out 50 to 100 frictions.

This criterion is definitely inapplicable to both erotically aroused women and the men who practice Taoism.

A man familiar with the "Tao of Love" and really loving his sex partner considers the third sex intercourse session even more joyful than the first one. The first copulation, similar to the first and only triggers the appetite for the main sex session. Experienced Taoism followers are eager to have the second and the third sex sessions. According to ancient Chinese, Taoism followers were always ready for prolongation of copulation as she likes, and this statement is even more applicable to females than males [9].

In Taoism followers, penile frictions greatly differ from the standard Western method. In beneficial circumstances and with an appropriate sex partner, Taoism followers are able to carry out the frictions with impressive endurance and energy. If the partners really impress each other and are well familiar with each other's bodies, they can reach an incredible degree of coordination. The duration, frequency and energy of individual sexual intercourse sessions should completely satisfy both sex partners. This is the only cause for Taoism followers to pay special attention to various types of thrusts (frictions).

Monotonous in-and-out frictions (thrusting) may render prolonged sex boring; however, it becomes advantageous, if a man is able to diversify mannery and poses.

The frictions carried out using varying velocities, forces and depths add to the range of produced nuances of pleasure and, thus, strengthen the love feelings of sexual partners. Also, these variations help a man to control his ejaculation and preserve penile erection for a sufficiently long period.

The method preferred of Tao *masters* and uniformly popular among women of all ages involves alternating use of 9 shallow *frictions* and 1 deep prolonged *love thrust*. According to Taoist *masters*, this is the best combination bringing maximum benefits to the couple. Women usually obtain much pleasure from this method: they are teased by unfulfilled hopes initially, but get satisfied finally.

Also, there are many other friction sequences, for example: the 9 shallow → 1 deep, or 3 shallow → 1 deep friction sequence.

The male sex partner or both the male and the female sex partners can choose the combinations and variations most appropriate for both of them. Also, a woman can intervene on her own initiative and lead this "sarabande dance" of their bodies. Both partners should obtain maximum pleasure; however, a man should preserve control over his emotions and feelings in order to prevent ejaculations from occurring too early or too frequently.

Mallanaga Vatsyayana describes the following friction variants [1]:

- ➤ The ordinary natural coitus - "drawing together";
- ➤ He moving his penis passed into the vagina in all directions using his hand - "knocking";
- ➤ He thrusting from above, with her pelvis tilted downward - "dagger";
- ➤ He quick switching down to opposite upward thrusts - "rubbing";
- ➤ He prolonged application of pressure after a thrust - "compressing";

185

- ➢ He impetuous forward movement of a man's lower part of the body after complete removal of penis from vagina - "rush attack";
- ➢ He targeted rubbing of only one side - "wild boar's thrust";
- ➢ He alternate rubbing of both sides - "ox's thrust";
- ➢ He permanently thrusts without removing the penis from the vagina before the cessation of desire - "sparrow's play".

All these variants should be used depending on the individual preferences of both sex partners.

... With his penis still in her vagina, let She have a rest, her forehead in contact with his forehead. When she restores her forces, resume your attacks...

The main signs of sexual desire in females include body relaxation, closing of the eyes, loss of shame and eagerness to unite coitally as tightly as possible. Her hands jerk, she sweats, bites him, does not let him to rise, pushes him with her leg and, after he has reached the acme of sexual pleasure, continues to act like a man" [1].

The signs of arousal in women described in the "Tao of Love" are as follows [9]:
- o She gazes at him, her breathing is hurried and difficult, and her voice is uncontrollably vibrating.
- o She closes her eyes, her nostrils dilate, and she loses the ability to speak.

- Her ears redden, her face flushes, but the tip of her tongue becomes slightly cooler.
- Her hands are hot, and her abdomen is warm; however, her speech becomes nearly inarticulate.
- Her face expresses a condition of fascination; her body is as soft as her abdomen; and her extremities weaken.
- Her mouth is dry under the tongue, and her body presses against his.
- Pulsation of external genital organs becomes more prominent; secretions are flowing, the Yin flow is running.

All the signs mentioned above confirm that she is highly aroused; however, he should continue to control the situation and gain mutual benefits from the intercourse without excessive haste.

One of the heroines of the antique erotic novel "Jou Pu Tuan" (English - "The Prayer Mat of Flesh") dating back to the period of the Ming Dynasty says: "You underestimate me. A man should shuttle back and forth one to two thousand times in a row to make me feel satisfied."

For somebody, this demand sounds like an indictment ☺, while others – enthusiasts and heroes – regard it as a generous proposal.

1300 years ago, Sun Su-Mo, a Taoist physician, formulated the following rule: "One hundred sex sessions should be carried our without ejaculation".

Some readers may object to this use of numbers, dimensions and quantities in relation to love on the grounds that they are more applicable to *sexual gyms*. Nowadays, however, all types of gymnastics have become popular, and, for a sensual woman, erotic gymnastics with her sex partner is much more pleasant than a monotonous treadmill. Therefore, comparing sexual records with sports records is wholly appropriate, and striving for excellence and acquiring potency are the integral elements of progress similar to sports.

If you are able to make 10 steps, very soon you will attain the ability to easily make 1000 steps in a row by adding one step each consecutive day.

And how exciting it is to win!

By the way, keep it easy; do not necessary to take everything too seriously...

Have you ever paid attention to any nonverbal signs of pleasure in women?

A man should identify the woman's unspoken wishes and learn how to fulfill them [9]:

- He is sleeping, but She opens her palms to play with his nephrite stem and envelops it with her hands. He notices changes in her breathing; it means that she wants him. His action is needed.

- Her eyes and eyebrows tremble, She lets out guttural sounds or utter incoherent words. It means that she is highly aroused; it is necessary to fulfill her wish. She wants him to enter her, her nostrils dilate, and her mouth opens.

- Her arms embtace his back, the lower part of her body moving. It means that She is highly aroused. She wants the *Yin flow* to run; her body trembles; and She presses tightly against his body. He has to continue the penetration and control himself of ejaculation.

- Her fragrant body becomes flaccid, her extremities are extended and immobile; she breathes with difficulty through her nose. It means that she wants the thrusts to resume. He has to identify the force and intensity of the thrusts that cause her to feel maximum pleasure and bring her to the peak, because this moment is already approaching.

- She holds her legs with her two hands and widely opens her nephrite gates. It means that she feels a lot of pleasure. He should continue without any decrease in intensity.

- Her tongue protrudes from the mouth like She is half-asleep or half-drunk. It means that her genital organs are waiting for both deep and shallow thrusts, and she wants them to be vigorous and diverse/multidirectional.

- She extends her legs and toes trying to retain his *nephrite stem* (penis) inside her, but is uncertain about the method he will use to satisfy her demand for thrusts and whispers something. It means that the *Yin flow* is running. Time loses its limits, and each thrust resonates in the whole body.

- Suddenly, she obtains the desired reward and slightly turns his body with her hand. She sweats a little and laughs. It means that she does not want him to stop the process (coitus).

- Sweet feeling has come already, and she is obtaining increasing pleasure; her *Yin flow* has come; she still holds him firmly. It means that her satisfaction is incomplete. She is eager to reach complete satisfaction, her sweating becomes copious.

- Her body becomes hot and moist due to sweat; her arms and legs are relaxed. If her discharges are complete, her body extends, and her eyes close like She is falling into a deep sleep. It shows that She is completely satisfied.

These details relating to time points separated by very short periods confirm that Taoist physicians examined this field very thoroughly; also, there are some evidences of the presence of a third person recording woman's reactions at each stage of sexual intercourse.

Possibly, some sex positions required a third participant, and such sex intercourse sessions could provide additional information.

Some detailed descriptions of female responses and corresponding male actions necessary for her to reach satisfaction are as follows:

o She holds him firmly with her both hands: it means that she wants the contact between the bodies to be closer.

o She rises her legs: it means that she wants his *nephrite stem* (vulva) to penetrate deeper.

o She contracts her abdominal muscles: it means that she wants the frictions to become shallower.

o Her thighs are moving: it means that she needs more frequent frictions and deeper penetration.

o She hooks him with her legs: it means that she wants deep frictions to become more frequent.

o She crosses her legs on his back: it means that she wants the penetration to become deeper and result in application of pressure to the uterine cervix.

o Her mons pubis presses against him and moves: it means that she is highly aroused and wants her clitoris to be contacted more tightly.

o She swings side-to-side: it means that she needs deep thrusts to the left and to the right.

o She rises her body to press against him, sometimes bending like a cat: it means that she feels extreme pleasure, but is not satisfied yet, and the coitus should be continued more intensively.

o She relaxes her body: it means that her body and extremities reached a quiet state.

o Her external genital organs have filled with fluid, her *Yin flow* began. He can see that she is happy.

Orgasm is the feeling of extreme sexual pleasure.

Orgasm (from Greek οργασμός orgasmos, from organ to mature, swell, also sexual climax) is the sudden discharge of accumulated sexual tension during the sexual response cycle, resulting in rhythmic muscular contractions in the pelvic region characterized by sexual pleasure.

It appears at the end of sexual intercourse and is accompanied by the feeling of sexual satisfactions and the relief from muscular tension and vasocongestion (i.e., hyperemia of genital organs). Studies have found increases in the hormone oxytocin at orgasm in both men and women. Oxytocin's role in increasing trust, pair bonding and reducing anxiety has meant it is sometimes referred to as the "love and trust" hormone.

Definitions of orgasm vary and there is sentiment that consensus on how to consistently classify it is absent. At least twenty-six definitions of orgasm were listed in the journal Clinical Psychology Review.

Orgasms may be achieved by a variety of activities. In men, sufficient penile stimulation can be achieved during vaginal or anal intercourse, oral sex (fellatio) or masturbation/non-penetrative sex. In women, sufficient sexual stimulation can be achieved during vaginal intercourse, oral sex (cunnilingus) or masturbation/non-penetrative sex. Orgasm may also be achieved by the use of a sex toy, such as a sensual vibrator or an erotic electrostimulation. It can additionally be achieved by stimulation of the nipples, uterus, or other erogenous zones, though this is rarer. In addition to physical stimulation, orgasm can be achieved from psychological arousal alone, such as during dreaming (nocturnal emission for males or females) or by orgasm control.

To a large extent, The "Tao of Love" appeared as a result of the need for provision of similar sexual satisfaction to both woman and man that existed in ancient China. When it was first formulated several thousand years ago, people understood this very well. During that epoch, it was known as The Tao of Yin and Yang Communication, the title emphasizing the harmony between the sexes.

To ensure the harmony between Yin (the female principle) and Yang (the male principle), a man should completely satisfy a woman.

The following three principles distinguish Taoism from other sex studies.

I.   According to the first principle, a man should determine the appropriate timing of ejaculation corresponding to his age and physical status. This preliminary planning should allow him to make love according to woman's wishes and ensure the duration and frequency of sex intercourse sessions that are necessary for her complete satisfaction.

II.   The second principle is radically contrary to the Western concepts of sex. Ancient Chinese thought that ejaculation, especially the uncontrolled one, is not the most ecstatic moment for men. A man armed with this knowledge is able to discover many other, more pleasant aspects of sex. On the other hand, it helps him to control ejaculation.

III. The third principle is the importance of woman's satisfaction. This principle has acquired wide recognition due to the works by Kinsey and consecutive works published by Western sexologists. In the recent years, importance of this principle has been additionally emphasized by various feminist movements, and, apparently, the validity of this principle will never be contested.

These principles formed the basis of the ancient Chinese philosophy of love. They not only helped men and women approach the goal of frequent and prolonged lovemaking corresponding to their wishes, but also enabled ancient China to benefit from sexual freedom and naturalness that persisted throughout the period of Taoism domination.

Taoists maintained that sexual harmony unites people with infinite forces of nature that, from their viewpoint, also have sexual features. Thus, the earth was related to the ideas of female, or Yin energy, while the sky corresponded to male, or Yang energy; interaction between them produces the Universe. Interaction between a man and a woman produces the Unity.

Also, there are other forms of gender unity.

According to one of existing beliefs, the Khajuraho temple erotic stone carving of a man and a woman embracing each other signify the union of Soul and God. The Indian Tantra schools that taught sex life rituals and symbols considered *yoga* (i.e., spiritual improvement) and *bhoga* (physical pleasure) as alternative ways leading to the same goal, the achievement of final freedom. These schools taught that controlled pleasure is the quickest and easiest way to soul salvation.

Tantric sex is one of the secret Tantric practices based on spiritual intimacy of sex partners. In the Tantric literature, the ritual Tantric sex is called *maithuna*, the ritual coitus. However, in contrast to the habitual viewpoint, *maithuna* is the final step of the five-step *pañcamakara* (i.e., "the five Ms") ritual ("M" is the first letter of the name for each step of the ritual).

The Tantric sex procedure implies the participant's ability to perceive the Life Energy concentrated in seven chakras and to voluntarily control it. This involves participant's ability to release the *Kundalini* energy stored in the lower chakra and to deliver it to the upper chakras located in the head via *sushumna*, the spinal energy channel. The Tantric sex procedure involves a prolonged sexual intercourse without ejaculation, because the latter is considered as the loss of male energy. Tantric sex is a coitus based on the concentration of all feelings. Before the sexual intercourse, the sex partners carry out a number of procedures to increase the sensitivity of smell and taste receptors.

Tantric ideas include:

- Considering sex as a self-improvement practice.
- Considering the partner as an embodied god.
- Considering each position as a yoga asana.
- Slow pace of the procedure ensuring gradual transition to each new status at higher energy levels.
- Making the pleasures more fine and aesthetic.
- Mental recognition of each instant with aim to "absorb" by own mind the energy released in the course of the Tantric sex procedure.
- Tantric sex procedure not always implies the usual sex completion, such as an orgasm; however, it ensures the possibility of "infinite development" of mind.

"In Tantrism, meditation should be the main element of the procedure, because it transforms the energy; however, its status has lowered to a secondary role. And many people who are sexually perverted and suppressed have joined the Tantra schools. They are not interested in any transformation; they want only to get rid of their sexual suppression and are interested only in sex... In the group, they are allowed to do everything they want; thus, they eliminate their suppression and feel liberatedness and lightness. However, this is a simple sexual orgy, not a Tantra experience... They should be allowed to get acquainted with Tantral feelings only after deep meditation. // Osho. *The Path of the Mystic.*

It's not an exaggeration to say that sex positions constitute the spice of love: in case of the absence of an infinite variety of sex positions, love would lose a large proportion of its ecstasy.

Possibly, you have noted that this chapter does not contain any descriptions of sex positions or erotic details of sexual intercourse. This is because there are no predetermined prescriptions, and each sex intercourse session is carried out ex tempore and depends on improvisation.

Vatsyayana Mallanaga describes the 64 sex positions, called "the arts" in Kama Sutra, the 8 methods for lovemaking, 8 positions in each method.

In Kama Sutra they are called "the 64 arts" and described in 64 chapters grouped into the following 7 sections:

- love in general, its place in human life (5 chapters);
- sexual intercourse (15 chapters), a section containing thorough analysis of kisses, various types of preliminary fondling and orgasms, a list of sex positions, oral sex (auparishtaka), paraphilias, as well as love triangle situations (i.e., relations between husband, wife and husband's mistress);

- relations with girls (9 chapters), courting and marriage;
- relations with married women (8 chapters), a section describing within-family relations between husband and wives and wives' behavior;
- relations with other men's wives (10 chapters), a section mostly dedicated to temptation;
- prostitution (9 chapters);
- secret instructions (6 chapters), a section dedicated to charming the surrounding people and restoring the desire [1].

The "Sexual intercourse" section describing sex practices and positions is the most well-known section of Kama Sutra, and its versions existing in the modern mass culture are quite different from the original version. Often, this section is even regarded as the whole book. However, sexual practices constitute only approximately one fifth of the book, and sex positions themselves are described in three chapters of the book probably consisting of no more than one hundred Sūtras in total. The rest of the sections describe relations between men and women. Kama Sutra calls sexual intercourse "the divine unity".

Vatsyayana thought that sex is not reprehensible, but frivolous attitude to sex is a sin [1].

In their instructions, the teachers characterize this knowledge as:

✓ Delighting

   ✓ Beneficial

      ✓ Leading to Goal

         ✓ Making happy

           ✓ Gorgeous!!!

For L O V E....

- satiation is dangerous,
- monotonousness is fatal...

It is important to determine, what all this means to you:

➢ simple movements,
➢ body play,
➢ muse of love.

If you have made 1000 *Steps* without reaching *BLISS*...

Something should be changed, for example:

✓ *the position,*
✓ *the music,*
✓ *the environment*
✓ *the condition*
✓ *the context*
✓ *your own attitude to what is happening ...*

         o   *smile and start again* ☺

✓ *turn this page*

P.S.

Global Orgasm (GORG) was an action originally scheduled for 22 December 2006 by an author and activist couple to coincide with the end of solstice. The idea was for participants throughout the world to have an orgasm during this one day while thinking about peace. Based on theories such as that of the noosphere and the work of the **Global Consciousness Project** at Princeton, there was indication that such an event would have a widespread positive effect on Human well-being. It has since become an annual event, and with participating Orgasms permitted to fall within a 24 hour period around the actual Solstice.

In the context of Directed Orgasm as a practice and habit, the Solstice day is used to culminate the practice in solidarity with others every year.

The Global Orgasm's organizers intended to use the phenomenon to create a positive, altruistic change in the noosphere of the Earth that can be measured by the Global Consciousness Project (GCP). The initial intention was that this might begin or contribute to a societal shift away from war, and towards peace, as a basis of foreign policy and cultural values.

GORGeous Feeling may be approving and empowering women, and changing perception about sexuality from "original sin" to "original blessing".

Sex is a most intimate form of dancing.

Leonard Luis Levinson, 1970

Salvador Dali. **"Nude-Couple-Large-Serpent"**

LOVE is the trust.

Merilyn Monroe

9. Psychoanalysis and the pleasure principle

The problem of nuances of female sexuality is still completely unclear. The terra incognita of female psychosexuality remains, to a large degree, unstudied; however, its discoverers were led by the psychoanalysis banner [Grigoriy Maraon, 1930].

All that belongs to the problem of "pleasure/displeasure" constitutes one of the most sensitive fields of modern psychology [22].

The term *psychoanalysis* was first used by Sigmund Freud in French on March 30, 1896; this term was used in his article about etiology of neuroses published in the Neurological Journal. In 1900, he published his first work "Interpretation of dreams" specially dedicated to analysis of neuroses on the basis of the results of evaluating the *night visions*. After his studies carried out with applying of free association methods, he concluded that, in the majority of *analysantes* (person who underwent of analysis), the neurosis stems from suppressed sexual desire (i.e., libido). In case of a libido development disorder – for example, mother-fixation (i.e., Oedipus complex) – libido can not be satisfied and manifests itself as symptoms of a psychic disease. Also, unsatisfied desire can be redirected to non-sex goals, this condition is called "*sublimation*". According to this concept, manifestations of suppressed sexual desire can be identified not only in dreams or neuroses, but also in literature and art or other products of human mind [21, 22].

In the early works of Freud, published before 1920, neuroses are considered to be the result of a conflict between the subconscious mind with its governing *"pleasure principle"* and the conscious mind with its governing *"reality principle"* aiming at self-preservation.

According to Freud, all the disorders result from human sexual life, and the leading role in their development is played by **the pleasure principle**.

In the psychoanalytical theory of mind we take it for granted that the course of mental processes is automatically regulated by the pleasure principle (the corresponding Freud's German term – *Lustprinzip*): that is to say, we believe that any given process originated in an unpleasant state of tension and there upon determines for itself such a path that its ultimate issue coincides with a relaxation of this tension, i.e., with avoidance of displeasure (German - *Unlust*) or with production of pleasure (German - *Lust*).

However, it is improper to say that the pleasure principle is the only factor that governs the course of psychic processes. If it were the only factor, the overwhelming majority of our mental processes would be accompanied by pleasure or lead to pleasure; however, all our usual experiences sharply contradict to this course of events. Therefore, the only possible conclusion is that the psychic processes are characterized by presence of a strong tendency to pleasure principle leadership, which tendency, however, is opposed by a variety of other factors and conditions, with the resulting outcome, accordingly, not always corresponding to the pleasure principle.

Analysis of this problem shows that **Eros** acts, since the very beginning of life, as a *"desire for life"*; it is opposed by the *"desire for death"* that appeared simultaneously with the appearance of organic life.

"Desire for life" is more related to our internal perceptions and acts like a peace disturber; it brings the tensions whose resolution is perceived as *pleasure*.

In short, it is quite probable that Eros' striving to include organic origins into all large unities acts as a replacement for the *"desire for improvement"* (however, we do not recognize the existence of such replacement). This, together with the effects of the *displacement*, could explain the phenomenon attributed to the latter.

At the beginning of mental life, striving for pleasure is more expressed and, simultaneously, more specific, than during the later stages; this should result in frequent local ruptures. During the period of higher maturity, the leadership of the pleasure principle is ensured to a much greater degree, but, similar to other desires, it fails to successfully circumvent the control mechanisms. *Sigmund Freud, 1920* [22].

In his later works, S. Freud concentrated on the conflict within the mental domain governed by the reality principle. In his work "The **Ego** and the **Id**", Freud distinguished the following three components of mind – the **Id** ("It"), the **Ego** ("Me") and the **Super-Ego** ("Super-Me"). **Id** represents the subconscious desires, and **Ego** represents the reality principle. **Super-Ego** is formed in the course of learning by the human being of the social norms that start to subconsciously govern the mind and cause the appearance of conscience and unconscious sense of guilt.

The majority of modern psychoanalysts recognize the suppressed sexuality as the source of *mental disorders, neuroses and affective reactions* [2, 11, 21, 22].

As a result of psychoanalysis sessions, in was found that the developing symptoms are the replacement for the strivings whose force is contributed to by the sources of sexual desire [21, 22, 26, 35]. This data totally complies with the existing knowledge related to the nature of psychoneuroses and hysterias and their risk factors. Hysterical character is defined by certain degree of abnormal sexual displacement, increased resistance against sexual desire due to the well-known mechanisms of *shame*, *disgust* and *morals* and quasi-instinctive avoidance of intellectual examination of the problem of sex resulting, in extreme cases, in complete ignorance about sex lasting until puberty.

Freud's notion of "*unconscious*" and method for evaluating the latent causes of symptoms, as well as his notion that mental processes are ensured by interaction between independent domains, form the basis of the majority of modern schools of psychoanalysis, psychotherapy and personality theory. The notion that works of art can be regarded as the result of neurotic feelings of their author and are the manifestations of deep unconscious mind caused a great effect on culture in the XXth century.

Hysteria, or "uterine fury" (the Ancient Greek word "hystera" means "uterus"), is an obsolete medical diagnosis currently corresponding to a number of mild or moderately severe mental disorders. The diagnosis "hysteria" first appeared in Ancient Greece and was described by Hyppocrates. His contemporary Plato described *uterine fury* as a condition that develops in case of a woman's inability to conceive. For a long time, this notion of the nature of hysteria had excluded the possibility of hysteria development in males. This term was used to describe specific malaise and disorders of behavior in females; for a long time, they had been thought to result from wandering of the uterus in the body - this hypothetical mechanism explains the name of this disorder. Hysteria is characterized by presence of demonstrative emotional reactions, including crying, laughing, screaming, convulsions, paralysis, loss of sensitivity, deafness, blindness, confusion, increased sexual activity, insomnia, irritableness, loss of appetite, loss of interest in sex and decreased libido, including *sexual anorexia* (absence of sexual appetite).

Since ancient Egyptian times, many gynecological diseases (their first description was found in the Kahun Medical Papyrus dating back to 1950 BC) had been considered as diseases of the uterus; though their descriptions did not contain disorders of behavioral or emotional, they mentioned "the treatment of the women diagnosed with *uterine spasms* who like to stay in bed".

Psychoanalysis eliminates the symptoms in hysterical patients using a hypothesis stating that these symptoms are a replacement (i.e., some kind of transcription) for a number of affective mental processes, wishes and strivings that, due to a special mental process called displacement, are precluded from elimination by conscious mental activity. These thoughts retained in the unconscious mind strive to be expressed in accordance with their affective power (i.e., to find a way out, or *Abf'uhr*) and, in hysterical patients, are released via conversion into somatic phenomena (i.e., into hysterical symptoms). The affective processes made conscious using an appropriate special symptom-reversal procedure provide the possibility to obtain the most accurate information on the nature and origin of these, previously unconscious, mental phenomena [22].

Marie Bonaparte (1882-1962), the Princess of Greece and Denmark, the great-granddaughter of Lucien Bonaparte, the Emperor Napoleone Bonaparte's brother, was the author of several books on female sexuality and, for a long time, an *analysante* of Sigmund Freud.

In 1907, Marie Bonaparte married Prince George of Greece, the Count of Corfu, and in 1909, at the age of 27, she gave birth to her son Pierre. She had high aspirations in connection with her marriage to the son of the Greek king; however, her intimate experience proved disappointing: she felt no languor or sexual ecstasy. Marital relations resembled a monotonous routine. When Marie Bonaparte learned her husband was latent homosexualist, she calmed down and started to look for partners for extrarelational sex. Industrialists, politicians, actors and elite hotel porters formed a sequence occasionally interrupted by French Prime Minister Aristide Briand who was well known as an experienced lover. However, the love extasy escaped her and, thus, became her idée fix.

Marie Bonaparte got acquainted with famous Paris and Vienna gynecologists and formed a group of women who reported to her their intimate experiences and problems. Having carried out these interviews and analysed the obtained data, she measured the distance between clitoris and vagina in more than 300 women using an ordinary ruler. It seemed to her that she had discovered a clue and formulated the so-called "rule of thumb" saying that a woman is unable to have an orgasm, if the distance between clitoris and vaginal vestibule exceeds the thumb width. Then, she measured the distance between her clitoris and vulva. The distance was found to be very large, about 3 centimeters. She started to look for a plastic surgeon to change her genital anatomical relationships. Being a princess generally recognized as a beauty, prosperous adult woman and two children's mother, she decided to ... move her clitor to another place with the only purpose of reaching a real sexual exstasy. She virtually forced Joseph Halbahn, a Vienna gynecologist, to carry out this surgery, and he operated on her genitals three times. However, this intervention failed to help her reach an orgasm. Marie was nearing suicide, but was saved by her acquaintance with S. Freud. She passed S. Freud's psychoanalysis sessions and became not only his favorite student, but his devoted follower as well. However, her longed-for orgasm was reached for the first time only 20 years later, at the age of 48 years.

It is interesting to get acquainted with both Marie Bonaparte's works and her personal feelings.

"Nature has never ensured perfect adaptation of organisms to their functions in the environment, and this thesis is especially clearly confirmed by insufficient adaptation of human beings to fulfilling their erotic function, this insufficiency being more expressed in females than in males. Frigidity and infertility are unrelated to each other.

...A man is unable to simulate erection or sperm ejaculation, the most reliable evidences of his pleasure. Despite the possibility to identify clitoral erection, woman's behavior provides very scanty clues for assessing the presence of her orgasm or degree of sexual satisfaction." Marie Bonaparte, 1921 [35].

Marie Bonaparte was the first to practice psychoanalysis in France without having medical education; nowadays, this is already regarded as a tradition, called *Dilettante Psychoanalysis*. Despite of a definite tendency to psychoanalysis "medicalisation" in some associations in the USA, psychoanalysis remains separate from psychotherapy and is an independent clinical practice all over the world, with medical or psychological education not being obligatory for beginning own analytical practice.

On November 4, 1926, Marie Bonaparte established the first, and, currently, the most influential, psychoanalytical society, the Psychoanalytical Society of Paris (Société Psychanalytique de Paris); and in 1927, at her own expense, she established and issued the first psychoanalytical journal in France, Revue Française de Psychanalyse. She attracted to France the leading psychoanalysts of her time, thus rendering Paris the world's center of psychoanalytical thought for many years ahead.

Marie Bonaparte had published the book "De la sexualite` de la femme" in 1951, France [35].

One of the most important topics discussed in this book was the masculinization of women: Marie Bonaparte predicted a decrease in the differences between sexes in the future. Since the very beginning, Marie Bonaparte had been an advocate of amateur (dilletant) psychoanalysis.

At the 20th International Psychoanalytical Congress in Paris (1957), Marie Bonaparte presented a report stating that "the half-century existence of psychoanalysis has resulted in liberation of sexuality, greater sexual freedom for women and better openness towards children. Mankind has become less hypocritical and, possibly, even more happy! The analysis helps to accept the reality of death and meet it more courageously, like Freud did it."

According to T. Parson's classical sex role theory, the anatomical characteristics of the body established during pregnancy automatically imply a definite model of conscious mind.

In his book "On character and libido development: six essays", Karl Abraham (1921) said: "Long time ago, we already expanded the scope of fundamental biogenetic law of organic evolution of man to cover the mental evolution of human beings.

According to Haeckel's biogenetic law, ontogenesis (i.e., individual development) shortly reproduces phylogenesis (i.e., historical development), but only during the embryonic stages of development [15].

Everyday psychoanalytical experience shows that each individual organism reproduces the species evolution also in the mental domain. The available extensive clinical experience allows us to formulate a special law specifically related to psychosexual evolution; according to this law, the psychosexual evolution always follows the organic (somatic) evolution and demonstrates some kind of a late repeat/reproduction of the same processes.

The biological processes mimicking the evolutionary processes occur during the most early embryonic period, whereas the psychosexual processes that we are studying occur in the course of a number of postpartum years, from birth to puberty. As regards embryology, there is a quite prominent parallelism between the stepwise psychosexual development observed by us and the evolutionary processes reflected during the early embryonic development" [35].

Both human and nonhuman (animal) organisms contain a definite program ensuring perpetuation of the species.

Use of the term **"libido"** to denote sexual desire was proposed by Freud (the Latin word *libido* means "desire", "sensuality"). Freud used this term in the wide sense; for example, he thought that libido also contributes to the acts of *defecating* and *sucking* in children.

We will understand this term as applying only to sex desire.

Libido includes a number of instinctive, involuntary desires that appear in combination with each other or individually depending on the age period.

Freud distinguished the following phases of libido evolution:

> *Oral* (i.e.., pregenital) phase

> *Anal* phase

> *Genital* phase.

According to Freud, a child whose libido is based on the most important somatic demands is characterized by presence of *oral* eroticism. During this period, his or her mother is the first object to interact with (so to say, a "pre-object"), and, initially, the child concentrating on his or her mother does not distinguish between her and himself/herself. During the first (autoerotic) phase characterized by presence of sucking activity, any differences between a girl and a boy as regards their behavior are absent [21, 22].

Sigmund Freud wrote:

"Masculinity implies a subject, activity and presence of penis; and femininity implies an object and passiveness (receptivity) [21].

216

S. Freud recognized the existence of only one gender, because he considered **female sexuality as initial stage of sexual development** [22].

According to S. Freud, the females that do not reject their masculinity and do not abandon their object of initial love and phallic stage erogenous zone (i.e., clitorocentrism), become lesbians.

In his work "Evolution of sexuality and intermediate conditions" (1930), Maraon, a Spanish biologist, stated on the basis of his clinical experience that each human being initially has the signs of both sexes, and, during later stages, one of the sexes, due to the effects of hormones, develops further and dominates, never suppressing, however, opposite sex manifestations.

According to Alfred Jost's (1972) law, signs of the male gender develops due to the virilizing effects of androgens.

If virilizing effects are absent, any individual organism, irrespective of his or her genetic gender, demonstrates feminine development [27, 28].

Female sexuality develops due to the effects of estrogens produced by the ovaries, provided that virilizing effects of androgens are absent. During later stages, when the effects of the ovaries decrease or disappear, female's hormonal status changes, and she can develop a psychological sexual crisis manifested by decreased libido, changed sexual preferences and modified erogenous feelings.

In his "Three essays on the theory of sexuality" (1905), S. Freud wrote: "Taking into consideration the autoerotic and masturbatory sexual manifestations, it can be supposed that small girl sexuality is quite masculine in character. These patterns make it possible to suppose that, during the period of yet unestablished masculinity lasting until the completion of sexual maturation, a male is subject to intergender crisis and is effeminate [21].

"As regards the application of the terms "masculine" or "feminine" to libido, it can be supposed that libido, by its nature, has masculine origin in both men and women and irrespective of its object (i.e., a man or a woman) [21].

Female orgasm is not a prerequisite for reproduction! In many cases, woman's erotic desire and, first of all, her ability to reach an orgasm increases with growing sexual experience [22].

In males, the orgasm actuator and the main erogenous-sensitive organ is the penis, a highly specialized organ abundantly supplied with blood vessels and nerve terminals. In females, the corresponding organ, the clitoris, remains in a rudimentary condition and, often, is weakly sensitive to insufficiently intensive or insufficiently long stimulation. Instead, female erogenous sensitivity spreads to adjacent vulvar, vaginal and anal mucosa and the whole skin, especially the skin of the breast region.

Psychoanalysts state that, in normal sexual intercourse, stable clitoral sensitivity and, in some cases, increased clitoral sensitivity are an obstacle to ensuring the necessary vaginal sensuality. In the women fighting against frigidity, the most valuable achievement is changing from *clitorocentric libido* and its peak, *the clitoral orgasm,* (i.e., the male power) to a vaginal erogenous sensitivity center (i.e., a purely feminine pathway) [21, 22, 35].

In her book "Psychoanalysis of the Sexual Functions of Women" (1925), Helene Deutsch stated that she observed many cases of the regression of female erogenous sensuality from vagina to clitoris after menopause. These women, previously satisfied during vaginal sexual intercourse, required additional fondling for reaching an orgasm.

Marie Bonaparte also supported the bisexual theory. She agreed with G. Maraon's theory maintaining that each human being initially has the signs of both sexes and stated that every human has both origins, masculine and feminine. At a later stage, one of the sexes dominate, but does not suppress the manifestations of the opposite sex. This bisexuality has significant consequences. In males, it is manifested by excessive flexibility of character; in females, its manifestations include masculinity complex and clitorocentrism.

Also, in females, there is a problem of conflict between the two existing erogenous zones. M. Bonaparte thought that vaginal orgasm exceeds the male orgasm. Therefore, the main problem is help for woman to adapt to her own erotic function.

Therefore, a woman able to reach vaginal orgasms has an advantage over a man, and women with increased vaginal ability easily reach deeper and more intensive orgasms. Clitoral super-sensitivity is a sign of the absence of adaptation and an intersexual phenomenon connected with bisexuality and a masculinity complex; it causes profound disorders of femininity in women.

Below are some thoughts of Marie Bonaparte that she presented in her book "De la sexualite' de la femme" (1951) as a psychologist and a woman who managed to change from sexual anesthesia and dissatisfaction to sexual freedom and harmony.

"All types of masochism are interrelated and have feminine origin – i.e., the wish to be beaten, whipped, penetrated and fertilized by a man.

A man should oppose the passive setting (and masochism in general) unless dictated by biology; and a woman should obey and accept it."

"In fact, whipping is an act that precedes penetration and rupture of a substance. They knock at the door before coming in.

"In an adult woman, vaginal sensuality during sexual intercourse is clearly based on a great masochist whipping fantasy and unconscious consent to the content of this fantasy... During sexual intercourse, a woman is really subjected to some kind of "whipping" with a man's penis. She accepts this "whipping" and often even likes it [35].

Examinations of vaginal mucosa sensitivity showed that vaginal mucosa is nearly insensitive to heating, cooling or pain stimulation. Intravaginal operations (for example, closing the vaginal ruptures) can be carried with nearly no anesthesia. Nevertheless, in an adult woman, vagina is an organ supplied with highly adapted and specialized erogenous sensitivity and is an origin of orgasm during sexual intercourse.

Possibly, this controversy stems from the following apparent controversies occurring in the course of woman's life since childhood: change in the object of love; change in the dominating erogenous zone; and change in the quality of sexual arousal.

Typical female orgasm starts not only due to penile head frictions against the vaginal or clitoral surface highly sensitive to these frictions, or due to diffuse sensitivity of vulvar mucosa that, of course, plays a role in the reaching of female orgasm together with the frequently encountered erotization of vaginal entrance circumference and posterior comissure of the pudendal lips. Female orgasm should be also contributed to by another type of sensitivity, the vaginal sensitivity; it should be deep and should respond to deep penile thrusts able to cause the tension of the bulb surrounding the vagina. The nature of vaginal erotic sensitivity should differ from that of the penile or clitoral head." [35].

Female masochism and its relation to frigidity

Women's vigorous protests against their masochism, passiveness (receptivity) and femininity have a solid bisexual basis connected with masculine manifestations. Without their masculinity, women would accept the natural female masochism completely and patiently.

Clitroral abandoning/"bankruptcy" results in the domination of vaginal erogenous function." [35].

"One psychologist who has consulted many lesbians said: This is definitely so. Although, the sexual attraction of lesbian relations, a girl with prevalent vaginal sensitivity always ends up with coming to a man who is more powerful in satisfying the erotism of her cavities and voids due to his penis." Only the women with predominantly or exclusively clitorocentric eroticism remain persistent lesbians. However, clitorocentric women often choose a male already at the early stages of their development, because clitoral erogenousness enables them to pursue both active and passive targets." [35].

Of course, sexual disposition of a woman can change: due to permanent variability, it has infinite number of variants. Evaluation of the main variants made it possible to divide clitorocentric women into the following two groups: the women confirming the clitoris and the women rejecting the vagina from the very beginning.

These two dispositions complement each other: rejection of the vagina implies confirmation of the clitoris and vice versa (a phenomenon called "hystery ban"); this mechanism prevents the development of frigidity. Prevalence of one of the two mutually complementing dispositions cause effect on the psychosexuality of the clitorocentric woman.

"In case of a normal sexual intercourse, with a clitorocentric woman in supine position and a man on top, the woman feels nearly nothing in view of her vaginal anesthesia. When discussing their frigidity, clitorocentric women complain of a too high location of the zone of sensitivity precluding it from being accessible for stimulation; they also express their wish to have this zone in a lower location accessible for stimulation during sexual intercourse in order to share the pleasure felt by their partner. A man can choose a position enabling him to carry out the frictions in the vagina and simultaneously stimulate the clitoris with his hand (or allow a woman to stimulate the clitoris with her own hand). This ensures the possibility for a woman to become completely satisfied. With a woman on top (for example, sitting on the thighs of her partner lying supine or on the thighs of her sitting partner), her clitoris directly contacts the root of the penis or the scrotum. In this case, additional manual stimulation of the clitoris can also be carried out by either man or woman. This results in a very different (i.e., more significant) sensual response on the part of the woman. In some cases, though it seems paradoxical at first sight, a clitorocentric woman, due to her strong protest against this "sadistic" penile penetration and her own masochism implied by such penetration, can remain unsatisfied notwithstanding the close contact between the penis and the clitoris ensured by this sex position.

Imagination of a vaginal woman during pleasurable penile frictions is supported, to some degree, by her unconscious notion of her vaginal cavity assuming the form of the passionately longed-for penis. Possibly, this mentality of vaginal women is opposite to clitorocentric women's mentality" [35].

The vaginism phenomenon is the local spasm occurring in case of the threat of sexual intercourse and aimed at preventing penile penetration. Vaginism can be transient or persistent (i.e., it can be an episode or a stable disorder in the erotic life of a woman). Psychological treatment is able to overcome this extreme reaction of female function rejection. Vaginism can be regarded as an extreme variant of *rejection of the vagina.*

In highly feminine women, erogenization and psychogenic receptiveness of the vagina and vaginal opening can be so strong that the penile penetration itself is able to bring much pleasure. This specific feature of the highly feminine women can be also revealed during both oral sex and anal penetration.

Psychologically, a woman performing fellatio (i.e., oral sex) can feel strong sexual arousal due to the sole fact of controlling the penis of the man she longed for and deep introduction of this penis. At the peak of pleasure, some women can feel sexual satisfaction and an orgasm.

In a woman subjected to a prolonged virginity loss process by her cautious lover, the hymen can be preserved due to its slow distension and then become the main erogenous zone. During each penetration, as well as during slow frictions, when the penile head (*corona penis*) passes through the vulvar ring, the hymen acts as an *erogenous string* to cause pleasure.

A lot of females have two zones – i.e, *the clitorocentric* zone and *the vulvovaginal* zone – ready for alternate stimulation since the very beginning of the sexual intercourse until the final pleasure. Some women who have got accustomed to masturbation before the beginning of vaginal sexual life become adapted to the vulvovaginal functioning and feel completely satisfied either with preliminary clitoral processing or without it immediately after defloration [26]. The majority of these women continue to masturbate after the beginning of regular vaginal sex; moreover, after the end of sexual intercourse, they feel the need to masturbate in order to experience *another* range of feelings that differ in intensity and emotional tinge. The fact of female post-coital masturbation should not be regarded by the woman's partner

225

as the sign of the absence of her satisfaction [23, 26]; on the contrary, it is a signal for initiating a new sex intercourse session in parallel with masturbation. A man is given a chance to observe the woman's self-satisfaction process (which is both pleasant to view and useful to know) and join it.

Obviously, erogenous zones are relatively independent from each other, because long-term female masturbation does not inhibit the vulvovaginal receptiveness.

It is thought that, in normal sexual intercourse, deeper orgasm is reached in case of stimulation of the region of the rectovaginal septum; however, but few women are able to precisely locate this zone of pleasure on the basis of their feelings. A man has better access to this zone and can apply rhythmic pressure to the region of posterior vaginal fornix and posterior vaginal wall in the vestibular region directly adjoining to the rectal wall using his penile head.

This chapter dedicated to psychology and psychoanalysis of female sexuality presents some thoughts and quotations in the form of dialogs. These thoughts and quotations belong to psychoanalytic Sigmund Freud; Freud's analysante and student Marie Bonaparte who experienced all the torments of sexual dissatisfaction; feminist Helene Deutsch; and sex therapy pioneer Helen Kaplan Singer; also, some thoughts are the result of this author's own gynecological observations. These dialogues are endless, which shows that no ideal or complete answers have been found.

Sexual energy moves people, occupies their minds, stimulates knowledge acquisition and contributes to development of relations and art...

In cases of abnormal, potentially dangerous, variants of constitutional predisposition, a sublimation process becomes possible; this process is characterized by switching the extremely powerful energy coming from individual sources of sexuality to other domains, with consecutive significant increase in the mental work ability. Sublimation process is one of the sources of artistic creativity: analysis of data about highly talented persons (especially those characterized by outstanding artistic abilities) reveals a variety of work ability, perversion and neurotic parameters that depend on the completeness of sublimation [21].

A special type of sublimation is the *inhibition* through *reaction formation*; it starts already in the latent period, in childhood, and, in favorable cases, continues throughout the lifespan [21, 22].

Salvador Dali (in Spanish: *Salvador Domènec Felip Jacint Dalí i Domènech, Marqués de Dalí de Púbol*; May 11, 1904 — January 23, 1989), a Spanish artist, one of the most famous surrealist artists.

In Salvador Dali, a thorough examiner of S. Freud's works, the unrealized libido and passion for women was converted into hyper-erogenous art images. This phenomenon of conversion of unrealized sexual desire and energy into artificial images (i.e., the realization of sexual neurotic energy in other deviant forms, including those involving art) was called by S. Freud the "libido sublimation". Salvador Dali was very impressed by Giacomo Casanova's short stories and Marquis de Sade novels. Inspired by Casanova's sensual revelations and intrigued by extraordinary personality of the scandalous author, Salvador Dali created illustrations that added savoir and eroticism to his short stories.

The artist's passions remain a subject of much conjecture. Also, it is not known whether these images are fruits of imagination or a reflection of reality.

Salvador Dali's "Surrealism is me!" was a reasonably substantiated statement.

Examination of the history of sexology clearly shows that, at present, development of medical technologies and general scientific and technological progress make it possible to study embryonic development, reproductive organ anatomy and specific features of erogenous zone innervation, as well as neurophysiology and hormonal regulation of sexuality, more profoundly. Drugs for treatment of anorgasmia and sexual anesthesia have appeared; methods for psychodynamic sex therapy have been developed; and various phalloimitators and vibrators are being widely used.

However, the following questions remain unanswered:

➤ Is woman's sexual satisfaction a prerequisite for reproduction (childbirth)?

➤ Has the sexual life quality improved?

Possibly, these questions will remain *rhetorical*..

Pathos - Humans are the most sexual species around!

Summing up, it should be noted that sexual energy always finds the way out! In favorable cases, this energy results in love manifestations, erotic feelings or artistic inspiration; usually, sexual energy results in various deviations…

Unrealized energy can result in neuroses, feeling of dissatisfaction or depression.

Releasing the powerful sexual energy, liberation of one's own mind and development of natural human instincts form the basis of inspiration, improving the health status and increasing life energy.

It is important to identify one's own erogenous strings, look at one's own body, release one's fantasy and look for natural desire!

Many questions are still waiting for your answers!

Most importantly, you will have to feel your own life, passions and muse…

Final Sting

Harmony of intimate relationship is possible when a female accepts a male with both her body and soul, while recognizing his masculine origin and phallic power.

Only then each touch and each movement can result in a storm of emotions and pleasure!

Then time loses its boundaries, fear of death fades away, and the feeling of eternity and bliss appears!

It means that the moment when you are happy together has come and will last forever!

And it is up to you to decide whether this is just lightning bliss of Eros...

or a cosmic sparkle of **procreation** initiating a new *LIFE*.

Salvador Dali. **"Girl-on-Rhinoceros-Horn"**

Epilogue

According to the main idea, in the beginning there was the word that created everything, and the word is the source of all powers. Because words consist of letters, each letter has its own power.

It is thought that the ancient alphabets (e.g., the Greek, Latin, Arab, Jewish and Chinese alphabets) were created using the laws of the universe, with each letter/hieroglyph, like an idea, having its power expressed in a number.

Thus, it is possible, by making letter combinations and creating words, to influence on the concealed world, discover new laws and predict future events.

Sex (six, sex) is associated with the number 6.

In numerology, the number 6 is the symbol for the planet Venus. It is the most perfect number within the 1 to 10 sequence; Philo of Alexandria called it "the most fertile of all numbers". It is the symbol of balance and harmony.

The symbol of the number 6 is the beehive, with its bees and honeycombs. The number 6 represents the harmony of the perfect enlivened matter; it is an ideal of primary life, its ideal embodiment. For Jews and Arabs, the sixth day of the week is the special day called Shabbat. In Kabbalah, the number 6 symbolizes 6 days of creation. In the Pythagorean system, the number 6 symbolizes success or luck (this sense has been preserved until present time in the dice); the same applies to the cube that has 6 faces and symbolizes stability and truth.

Σ, σ, ς (sigma; the Greek word: σίγμα), the 18th letter of the Greek alphabet, corresponds to the number 200. The Latin letter S and the Cyrillic letter C originated from the letter "sigma"; in the books, the letter S was printed as **ς**.

Author's logotype on the cover page:

✓ the first capital letter "sigma" (**Ϭ**) visually corresponds to the number 6 and contains the male vector symbol and the Eros arrow ♂.

✓ the number 6 is the symbol of sexuality, love and harmony.

✓ the second letter **ς** is the mirror of Venus ♀, a symbol of femininity.

✓ the capital letter **Ϭ** and the last letter **ς** symbolize the merging of male and female forces; combination of the two letters is read as **sex** (transcription: [SX]).

✓ the center of merging of the two letters **σ** + **ς** corresponds to a symbol of eternity and harmony of male and female forces.

✓ in the Armenian cuneiform script, spiral is the sign of the sun and symbol of eternity.

✓ the oval in the center is the symbol of the origin of life; the oval is surrounded by a contour of embryo.

References

1. Kamasutra by Mallanaga Vatsyayana, translated by Wendy Doniger, Oxford University Press 2003, ISBN 978-0-19-283982-4.
2. Бухановский А.О., Кутявин Ю.А., Литвак М.Е. Общая психопатология. Пособие для врачей. Ростов н/Д.: Изд-во ЛРНЦ «Феникс», 1998. — с. 416.
3. Von Prader A. Der genitalbefund beim pseudohermaproditus feminus des kongenitalen adrenogenitalen syndromes. Helv Pediatr Acta 1954; 9: 231–248.
4. Иван Блох. История проституции, перевод с немецкого Санкт-Петербург, Медицина, 1913.
5. Васильченко Г.С. Патогенетические механизмы импотенции. Медгиз, 1956.
6. Sadler T.W. Langman`s Medical Embryology. Williams&Wilkins USA, 2000; p. 215.
7. Карлсон Б. Основы эмбриологии по Пэттену. В двух томах. М.: Мир 1983; с. 367; 389.
8. Мильман Л.Я. Импотенция. Ленинград, изд. Медицина, 1965.
9. Жолань Чжан. Дао Любви, Перевод В. Зайцевой, изд. Советский спорт, ISBN 5-85-009-341-9; 1993.
10. Павлов И.П. Полное собрание сочинений. М.-Л., изд. АН СССР, 1951, том IV, с. 25-26.
11. Свядощ А.М. Неврозы и их лечение. Медгиз, 1959.
12. Якобзон Л.Я. Половое воздержание перед судом медицины. Ленинград, Медицина, 1906.
13. Якобзон Л.Я. Половая холодность женщины. Ленинград, Медицина, 1927.
14. Сапин М.Р. Анатомия человека в 2 томах, Москва, Медицина, 1986.
15. Биология // под редакцией В.Н. Ярыгина, Москва, Медицина, 1985.
16. Гистология // под редакцией Ю.И. Афанасьева, Н.А. Юриной, Москва, Медицина 1989.
17. Макиян З.Н. Аномалии женских половых органов: систематизация и тактика оперативного лечения. Дисс. на доктора мед. наук, Москва, 2011. // Makiyan Z. Anomaly of female genitalia: systematization and surgical treatment. Doctor of medicine Dissertation, Moscow, Russia, 2011

18. Makiyan Z. Anomaly of sexual development (Russian edition) // Макиян З.Н. Аномалии развития пола: методы оперативного лечения в гинекологии //, Lambert Academic Publishing, Saarbrucken, Germany, 2012. ISBN - 978-3-8473-2664-9.
19. Мандельштам А.Э. Семиотика и диагностика женских болезней, Москва, Медицина, 1976.
20. Кон И. С. Введение в сексологию. Курс лекций. Учебное пособие для вузов. — М.: Олимп, Инфра-М, 1999. — 288 с. - ISBN 5-86225-624-5
21. Sigmund Freud. "Three essays on the theory of sexuality", 1905.
22. Sigmund Freud. " Beyond the Pleasure Principle", 1920.
23. Alfred Charles Kinsey. Sexual Behavior in the Human Male, USA, 1948.
24. Alfred Charles Kinsey. Sexual Behavior in the Human Female, USA, 1953.
25. Masters, William H. and Johnson, Virginia E. Human Sexual Response. — No, 1966. — ISBN 0-316-54987-8
26. Helen Kaplan Singer. The New Sex Therapy: Active Treatment of Sexual Dysfunctions. New York: Routledge, 1974
27. Jost A., Vigier B., Prepin J. 1972 Freemartins in cattle: the first steps of sexual organogenesis. J Reprod Fertil 29:349–79
28. Jost A. A new look at the mechanism controlling sex differentiation in mammals. Johns Hopkins Medical Journal, 1972, 130, p.28-36.
29. Acien P., Acien M., Sanchez-Ferrer M. Complex malformations of the female genital tract. New types and revision of classification.// Human Reproduction October 2004 19(10).
30. Peter A. Lee, Christopher P. Houk, S. Faisal Ahmed. Consensus Statement on Management of Intersex Disorders.// PEDIATRICS Vol. 118 No. 2 August 2006, pp. e488-e500
31. Brody Stuart, Weiss Petr. Vaginal orgasm is associated with vaginal (not clitoral) sex education, focusing mental attention on vaginal sensations, intercourse duration, and a preference for a longer penis // The Journal of Sexual Medicine. Published online. September 2009.
32. Komisaruk BR, Wise N, Frangos E, Liu W-C, Allen K, and Brody S. Women's clitoris, vagina and cervix mapped on the sensory cortex: fMRI evidence. J Sex Med 2011;8:2822–2830.
33. Rowland DL. Neurobiology of sexual response in men and women. CNS Spectr. 2006 Aug;11(8 Suppl 9):6-12.

34. Brotto LA, Bitzer J, Laan E, Leiblum S, Luria M. Women's sexual desire and arousal disorders. J Sex Med. 2010 Jan;7(1 Pt 2):586-614. Review. Erratum in: J Sex Med. 2010 Feb;7(2 Pt 1):856.

35. Marie Bonaparte. De la sexualite` de la femme, 1951.

36. Leiblum SR. Definition and classification of female sexual disorders. Int J Impot Res 1998;10:S104-6.

37. Bozman AW, Beck JG. Covariation of sexual desire and sexual arousal: The effects of anger and anxiety. Arch Sex Behav 1991;20:47-60.

38. Slob AK, Bax CM, Hop WCJ, Rowland DL, van der Werff ten Bosch JJ. Sexual arousability and the menstrual cycle. Psychoneuroendocrinology 1996;2:545-58.

39. Sanders SA, Graham CA, Milhausen RR. Predicting sexual problems in women: The relevance of sexual excitation and sexual inhibition. Arch Sex Behav 2008;37:241-51.

40. Brotto LA, Heiman JR, Tolman D. Narratives of desire in mid-age women with and without arousal difficulties. J Sex Res 2009;46:387-98.

41. Graham CA, Sanders SA, Milhausen RR, McBride KR. Turning on and turning off: A focus group study of the factors that affect women's sexual arousal. Arch Sex Behav 2004;33:527-38.

42. Hartmann U, Heiser K, Rьffer-Hesse C, Kloth G. Female sexual desire disorders: Subtypes, classification, personality factors and new directions for treatment. World J Urol 2002;20:79-88.

43. Sand M, Fisher WA. Women's endorsement of models of female sexual response: The nurses' sexuality study. J Sex Med 2007;4:708-19.

44. Brotto LA. The DSM diagnostic criteria for hypoactive sexual desire disorder. Arch Sex Behav DOI:1007/S10508-009-9543-1. American Psychiatric Association. Diagnostic and statistical manual of mental disorders, 4th edition. Text Revision. Washington DC: American Psychiatric Association; 2000.

45. Basson R, Leiblum S, Brotto L, Derogatis L, Fourcroy J, Fugl-Meyer K, Graziottin A, Heiman JR, Laan E, Meston C, Schover L, van Lankveld J, Weijmar Schultz W. Definitions of women's sexual dysfunction reconsidered: Advocating expansion and revision. J Psychosom Obstet Gynaecol 2003; 24:221-9.

46. Cain VS, Johannes CB, Avis NE, Mohr B, Schocken M, Skurnick J, Ory M. Sexual functioning and practices in a multi-

ethnic study of midlife women: Baseline results from SWAN. J Sex Res 2003;40:266–76.

47. Rosen RC, Shifren JL, Monz BU, Odom DM, Russo PA, Johannes CB. Correlates of sexually related personal distress in women with low sexual desire. J Sex Med 2009;6:1549–60.

48. Dunn KM, Croft PR, Hackett GI. Satisfaction in the sex life of a general population sample. J Sex Marital Ther 2000;26:141–51.

49. King M, Holt V, Nazareth I. Women's views of their sexual difficulties: Agreement and disagreement with clinical diagnoses. Arch Sex Behav 2007;36:281–8.

50. Both S, Everaerd W, Laan E. Modulation of spinal reflexes by aversive and sexually appetitive stimuli. Psychophysiology 2003;40:174–83.

51. Both S, Spiering M, Everaerd W, Laan E. Sexual behavior and responsiveness to sexual stimuli following laboratory-induced sexual arousal. J Sex Res 2004;41:242–58.

52. Everaerd W, Laan E. Desire for passion: Energetics of sexual response. J Sex Marital Ther 1995;21:255–63.

53. Laan E, Everaerd W. Determinants of female sexual arousal: Psychophysiological theory and data. Annu Rev Sex Res 1995;6:32–76.

54. Laan E, Everaerd W, van der Velde J, Geer JH. Determinants of subjective experience of sexual arousal in women: Feedback from genital arousal and erotic stimulus content. Psychophysiology 1995;32:444–51.

55. Ефрон, Илья Абрамович // Еврейская энциклопедия Брокгауза и Ефрона. Санкт-Петербург, 1906—1913.

56. McCall K, Meston C. Differences between preand postmenopausal women in cues for sexual desire. J Sex Med 2007;4:364–71.

57. Meston CM, Buss DM. Why humans have sex. Arch Sex Behav 2007;36:477–507.

58. Basson R. Using a different model for female sexual response to address women's problematic low sexual desire. J Sex Marital Ther 2001;27:395–403.

59. Basson R, Brotto LA. Management of low sexual desire in women. In: Balon R, Segraves RT, eds. Clinical manual of sexual disorders. Arlington, VI:American Psychiatric Publishing; 2009:119–59.

60. Giles KR, McCabe MP. Conceptualising women's sexual function: Linear vs. circular models of sexual response. J Sex Med 2009;6:2761–71.

61. Oberg K, Fugl-Meyer AR, Fugl-Meyer KS. On categorization and quantification of women's sexual dysfunctions: An epidemiological approach. Int J Impot Res 2004;16:261–9.

62. Shifren JL, Monz BU, Russo PA, Segreti A, Johannes CB. Sexual problems and distress in United States women: Prevalence and correlates. Obstet Gynecol 2008;112:970–8.

63. Bancroft J, Loftus J, Long JS. Distress about sex: A national survey of women in heterosexual relationships. Arch Sex Behav 2003;32:193–208.

64. Nathan SG. When do we say a woman's sexuality is dysfunctional? In: Levine SB, Risen CB, Althof SE, eds. Handbook of clinical sexuality for mental health professionals. New York: Brunner-Routledge; 2003:95–110.

65. Chivers M, Seto M, Lalumiere M, Laan E, Grimbos T. Agreement of genital and subjective measures of sexual arousal: A meta-analysis. Arch Sex Behav In Press.

66. Laan E, Janssen E. How do men and women feel:Determinants of subjective experience of sexual arousal. In: Janssen E, ed. The psychophysiology of sex. Bloomington, IN: Indiana University Press; 2007:278–90.

67. Heiman JR, Meston CM. Empirically validated treatment for sexual dysfunction. Ann Rev Sex Res 1997;8:148–94.

68. Brotto LA, Basson R, Gorzalka BB. Psychophysiological assessment in premenopausal sexual arousal disorder. J Sex Med 2004;1:266–77.

69. Graham CA. The DSM diagnostic criteria for female sexual arousal disorder. Arch Sex Behav 2009; Sep 24 [Epub ahead of print] doi:10.1007/S/0508-009-9535-1.

70. Graham CA. The DSM diagnostic criterial for female orgasmic disorder. Arch Sex Behav 2009; Sep 26 [Epub ahead of print] doi: 10.1007/S10508-009-9542-2.

71. Laumann EO, Paik A, Rosen R. Sexual dysfunction in the United States: Prevalence and predictors. JAMA 1999;281:537–44.

72. Fugl-Meyer AR, Sjogren Fugl-Meyer K. Sexual disabilities, problems and satisfaction in 18–74 years old Swedes. Scand J Sex 1999;2:79–105.

73. Mercer CH, Fenton KA, Johnson AM,Wellings K, Macdowall W, McManus S, Nanchahal K, Erens B. Sexual function

problems and help seeking behaviour in Britain: National probability sample survey. BMJ 2003;327:426–7.

74. Laumann EO, Nicolosi A, Glasser DB, Paik A, Gingell C, Moreira E, Wang T. Sexual problems among women and men aged 40–80 years: Prevalence and correlates identified in the Global Study of Sexual Attitudes and Behaviors. Int J Impotence Res 2005;17:39–57.

75. Leiblum SR, Koochaki PE, Rodenberg CA, Barton IP, Rosen RC. Hypoactive sexual desire disorder in postmenopausal women: US results from the Women's International Study of Health and Sexuality (WISHeS). Menopause 2006;13:46–56.

76. Dennerstein L, Koochaki P, Barton I, Graziottin A. Hypoactive sexual desire disorder in menopausal women: A survey of western European women. J Sex Med 2006;3:212–22.

77. West SL, D'Aloisio AA, Agans RP, Kalsbeek WD, Borisov NN, Thorp JM. Prevalence of low sexual desire and hypoactive sexual desire disorder in a nationally representative sample of US women. Arch Intern Med 2008;168:1441–9.

78. Witting K, Santtila P, Varjonen M, Jern P, Johansson A, von der Pahlen B, Sandnabba K. Female sexual dysfunction, sexual distress, and compatibility with partner. J Sex Med 2008;5:2587–99.

79. Hayes RD. Assessing female sexual dysfunction in epidemiological studies: Why is it necessary to measure both low sexual function and sexuallyrelated distress? Sex Health 2008;5:215–8.

80. Dunn KM, Croft PR, Hackett GI. Sexual problems: A study of the prevalence and need for health care in the general population. Fam Pract 1998;15: 519–24.

81. Basson R. Rethinking low sexual desire in women. Br J Obstet Gynaecol 2002;109:357–63.

82. Basson R. Biopsychosocial models of women's sexual response: Applications to management of desire disorders. Sex Relat Ther 2003;18:107–15.

83. Meston CM, Trapnell P. Development and validation of a five-factor sexual satisfaction and distress scale for women: The Sexual Satisfaction Scale for Women (SSS-W). J Sex Med 2005;2:66–81.

84. Phillippsohn S, Hartmann U. Determinants of sexual satisfaction in a sample of German women. J Sex Med 2009;6:1001–10.

85. Byers ES, Macneil S. Further validation of the interpersonal exchange model of sexual satisfaction. J Sex Marital Ther 2006;32:53–69.
86. Lutfey K, Link C, Rosen R, Wiegel M, McKinlay J. Prevalence and correlates of sexual activity and function in women: Results from the Boston Area Community Health (BACH) survey. Arch Sex Behav 2009;38:514–27.
87. Damasio A. Looking for Spinoza: Joy, sorrow, and the feeling brain. Orlando, FL: Harcourt; 2003.
88. Laan E, Both S. What makes women experience desire? Fem Psychol 2008;18:505–14.
89. Dennerstein L, Alexander JL, Graziottin A. Sexual desire disorder in women. In: Porst H, Buvat J, eds. Standard practice in sexual medicine. Oxford: Blackwell Publishing; 2006:315–9.
90. Bitzer J, Alder J. Sexuality during pregnancy and the postpartum period. J Sex Educ Ther 2000;25: 49–58.
91. Warnock JK, Clayton A, Croft H, Segraves R, Biggs FC. Comparison of androgens in women with hypoactive sexual desire disorder: Those on combined oral contraceptives (COCs) vs. those not on COCs. J Sex Med 2006;3:878–82.
92. Bancroft J, Hammond G, Graham C. Do oral contraceptives produce irreversible effects on women's sexuality? J Sex Med 2006;3:567.
93. Everaerd W, Both S, Laan E. The experience of sexual emotions. Ann Rev Sex Res 2006;17:183–99.
94. Leonard LM, Follette VM. Sexual functioning in women reporting a history of child sexual abuse: Review of the empirical literature and clinical implications. Ann Rev Sex Res 2002;13:346–88.
95. Fergusson DM, Mullen PE. Childhood sexual abuse: An evidence based perspective. Thousand Oaks, CA: Sage; 1999.
96. Meston CM, Heiman JR. Sexual abuse and sexual function: Examination of sexually relevant cognitive processes. J Consult Clin Psychol 2000;68: 399–406.
97. Oberg K, Fugl-Meyer K, Fugl-Meyer A. On sexual well-being in sexually abused Swedish women: Epidemiological aspects. Sex Relat Ther 2002;17: 329–41.
98. van Berlo W, Ensink B. Problems with sexuality after sexual assault. Ann Rev Sex Res 2000;11: 235–57.
99. Levin RJ, van Berlo W. Sexual arousal and orgasm in subjects who experience forced or non- J Sex Med

2010;7:586–614 consensual sexual stimulation—A review. J Clin Forensic Med 2004;11:82–8.

100. Rellini AH, Meston CM. Sexual desire and linguistic analysis: A comparison of sexually-abused and non-abused women. Arch Sex Behav 2007;36:67– 77.

101. Elmerstig E, Wijma B, Swahnberg K. Young Swedish women's experience of pain and discomfort during sexual intercourse. Acta Obstet Gynecol Scand 2009;88:98–103.

102. Cranston-Cuebas MA, Barlow DH. Cognitive and affective contributions to sexual functioning. Ann Rev Sex Res 1990;1:119–61.

103. Kuile MM, Vigeveno D, Laan E. Preliminary evidence that acute and chronic daily psychological stress affect sexual arousal in sexually functional women. Behav Res Ther 2007;45:2078–89.

104. Salemink E, van Lankveld JJ. The effects of increasing neutral distraction on sexual responding of women with and without sexual problems. Arch Sex Behav 2006;35:175–86.

105. Adams AE, III, Haynes SN, Brayer MA. Cognitive distraction in female sexual arousal. Psychophysiology 1985;22:689–96.

106. Dove NL,Wiederman MW. Cognitive distraction and women's sexual functioning. J Sex Marital Ther 2000;26:67–78.

107. van Lankveld J, Bergh S. The interaction of state and trait aspects of self-focused attention affects genital, but not subjective, sexual arousal in sexually functional women. Behav Res Ther 2008;46:514–28.

108. Barlow DH. Causes of sexual dysfunction: The role of anxiety and cognitive interference. J Consult Clin Psychol 1986;54:140–8.

109. Norton GR, Jehu D. The role of anxiety in sexual dysfunctions: A review. Arch Sex Behav 1984;13: 165–83.

110. Cooper AJ. Some personality factors in frigidity. J Psychosom Res 1969;13:149–55.

111. DeRogatis LR, Meyer JK. A psychological profile of the sexual dysfunction. Arch Sex Behav 1979; 8:201–23.

112. Kaplan HS. The sexual desire disorders. New York: Brunner & Mazel; 1995

113. Campillo GG, Bravo CS, Carmona FM, Perales RD, Calderon AV. Anxiety and depression levels in women with and without sexual disorders: A comparative study. Rev Mex Psicol 1999;16:17–23.

114. Trudel G, Landry L, Larose L. Low sexual desire: The role of anxiety, depression and marital adjustment. Sex Mar Ther 1997;12:95–9.
115. Bartlik B, Kocsis JH, Legere R, Villaluz J, Kosoy A, Gelenberg AJ. Sexual dysfunction secondary to depressive disorders. J Gend Specif Med 1999;2:52–60.
116. Apt C, Hurlbert DF. The sexual attitudes, behavior, and relationships of women with histrionic personality disorder. J Sex Marital Ther 1994;20: 125–33.
117. Wiederman MW. Women's body image selfconsciousness during physical intimacy with a partner. J Sex Res 2000;37:60–8.
118. Dean J, Rubio-Aurioles E, McCabe M, Eardley I, Speakman M, Buvat J, Tejada ISD, Fisher W. Integrating partners into erectile dysfunction treatment: Improving the sexual experience for the couple. Int J Clin Pract 2008;62:127–33.
119. Rubio-Aurioles E, Kim ED, Rosen RC, Porst H, Burns P, Zeigler H,Wong DG. Impact on erectile function and sexual quality of life of couples: A double-blind, randomized, placebo-controlled trial of tadalafil taken once daily. J Sex Med 2009; 6:1314–23.
120. Klusmann D. Sexual motivation and the duration of partnership. Arch Sex Behav 2002;31:275–87.
121. Laan E, van Driel EM, van Lunsen RHW. Genital responsiveness in healthy women with and without sexual arousal disorder. J Sex Med 2008;5:1424–35.
122. Rosen R, Brown C, Heiman J, Leiblum S, Meston C, Shabsigh R, Fergson D, D'Agrostino R. The Female Sexual Function Index (FSFI): A multidimensional self-report instrument for the assessment of female sexual function. J Sex Marital Ther 2000;26:191–208.
123. Graziottin A, Leiblum S. Biological and psychosocial pathophysiology of female sexual dysfunction during the menopause transition. J Sex Med 2005;2(3 suppl):S133–45.
124. Rust J, Golombok S. The GRISS: A psychometric instrument for the assessment of sexual dysfunction. Arch Sex Behav 1986;15:157–65.
125. Taylor JF, Rosen RC, Leiblum SR. Self-report assessment of female sexual function: Psychometric evaluation of the brief index of sexual functioning for women. Arch Sex Behav 1994;23:627–43.

126. Spector IP, Carey MP, Steinberg L. The sexual desire inventory: Development, factor structure, and evidence of reliability. J Sex Marital Ther 1996;22:175–90.

127. Derogatis LR. The Derogatis Interview for Sexual Functioning (DISF/DISF-SR): An introductory report. J Sex Marital Ther 1997;23:291–304.

128. Meyer-Bahlburg HFL, Dolezal C. The Female Sexual Function Index: A methodological critique and suggestions for improvement. J Sex Marital Ther 2007;33:217–24.

129. Brotto LA. Letter to the editor. J Sex Marital Ther 2009;35:161–3.

130. Quirk FH, Heiman JR, Rosen RC, Laan E, Smith MD, Boolell M. Development of a sexual function questionnaire for clinical trials of female sexual dysfunction. J Womens Health Gend Based Med 2004;11:277–89.

131. DeRogatis LR, Rosen R, Leiblum S, Burnett A, Heiman J. The Female Sexual Distress Scale (FSDS): Initial validation of a standardized scale J Sex Med 2010;7:586–614

132. Clayton AH, Segraves RT, Leiblum S, Basson R, Pyke R, Cotton D, Lewis-D'Agostino D, Evans, KR, Sills TL, Wunderlich GR. Reliability and validity of the Sexual Interest and Desire Inventory-Female (SIDI-F), a scale designed to measure severity of female hypoactive sexual desire disorder. J Sex Marital Ther 2006;32:115–35.

133. Leiblum S, Symonds T, Moore J, Soni P, Steinberg S, Sisson M. A methodology study to develop and validate a screener for hypoactive sexual desire disorder in postmenopausal women. J Sex Med 2006;3:455–64.

134. DeRogatis L, Clayton A, Lewis-D'Agostino D, Wunderlich G, Fu Y. Validation of the female sexual distress scale-revised for assessing distress in women with hypoactive sexual desire disorder. J Sex Med 2008;5:357–64.

135. Clayton AH, Goldfischer ER, Goldstein I, DeRogatis L, Lewis-D'Agostino DJ, Pyke R. Validation of the Decreased Sexual Desire Screener (DSDS): A brief diagnostic instrument for generalized acquired female hypoactive sexual desire disorder (HSDD). J Sex Med 2009;6:730–8.

136. DeRogatis LR, Allgood A, Rosen RC, Leiblum S, Zipfel L, Guo C. Development and evaluation of the Women's Sexual Interest Diagnostic Interview (WSID): A structured interview to diagnose hypoactive sexual desire disorder (HSDD) in standardized patients. J Sex Med 2008;5:2827–41.

137. Althof SE, Dean J, Derogatis LR, Rosen RC, Sisson M. Current perspectives on the clinical assessment and diagnosis of female sexual dysfunction and clinical studies of potential therapies: A statement of concern. J Sex Med 2005;2:147–54.
138. Sintchak G, Geer JH. A vaginal plethysmograph system. Psychophysiology 1975;12:113–5.
139. Levin RJ. The mechanisms of human female sexual arousal. Ann Rev Sex Res 1992;3:1–48.
140. Wouda JC, Hartman PM, Bakker RM, Bakker JO, van de Wiel HBB, Weijmar Schultz WCM. Vaginal plethysmography in women with dyspareunia. J Sex Res 1998;35:141–7.
141. Brauer M, Laan E, ter Kuile M. Sexual arousal in women with superficial dyspareunia. Arch Sex Behav 2006;35:187–96.
142. Brauer M, ter Kuile MM, Janssen SA, Laan E. The effect of pain-related fear on sexual arousal in women with superficial dyspareunia. Eur J Pain 2007;11:788–98.
143. Brauer M, ter Kuile MM, Laan E, Trimbos B. Cognitive-affective correlates and predictors of superficial dyspareunia. J Sex Marital Ther 2009; 35:1–24.
144. Meston CM, Gorzalka BB. Differential effects of sympathetic activation on sexual arousal in sexually dysfunctional and functional women. J Abnorm Psychol 1996;105:582–91.
145. Morokoff PJ, Heiman JR. Effects of erotic stimuli on sexually functional and dysfunctional women: Multiple measures before and after sex therapy. Behav Res Ther 1980;18:127–37.
146. Rellini A, Meston C. The sensitivity of event logs, self-administered questionnaires and photoplethysmography to detect treatment-induced changes in female sexual arousal disorder (FSAD) diagnosis. J Sex Med 2006;3:283–91.
147. Wagner G, Levin R. Oxygen tension of the vaginal surface during sexual stimulation in the human. Fertil Steril 1978;30:50–3.
148. Wagner G, Levin RJ. Effect of atropine and methylatropine on human vaginal blood flow, sexual arousal and climax. Acta Pharmacol Toxicol 1980;46:321–5.
149. Wagner G, Ottesen B. Vaginal blood flow during sexual stimulation. Obstet Gynecol 1980;56:621–4.
150. Hoon PW, Coleman E, Amberson J, Ling F. A possible physiological marker of female sexual dysfunction. Biol Psychiatry 1981;16:1101–5.

151. Levin RJ, Wagner G. Sexual arousal in women: Which haemodynamic measure gives the best assessment? J Physiol 1980;392:22–3P.
152. Slob AK, Ernste M, van der Werff ten Bosch JJ. Menstrual cycle phase and sexual arousability in women. Arch Sex Behav 1991;20:567–77.
153. Slob AK, Koster J, Radder JK, van der Werff ten Bosch JJ. Sexuality and psychophysiological functioning in women with diabetes mellitus. J Sex Marital Ther 1990;16:59–69.
154. Payne K, Binik Y. Reviving the labial thermistor clip. Arch Sex Behav 2006;35:111–3.
155. Prause N, Heiman JR. Assessing female sexual arousal with the labial thermistor: Response specificity and construct validity. Int J Psychophysiol 2009;72:115–22.
156. Henson DE, Rubin HB, Henson C. Labial and vaginal blood volume responses to visual and tactile stimuli. Arch Sex Behav 1982;11:23–31.
157. Prause N, Janssen E. Blood flow: Vaginal photoplethysmography. In: Goldstein I, Meston CM, Davis SR, Traish AM, eds. Women's sexual function and dysfunction: Study, diagnosis and treatment. London: Taylor & Francis; 2006:359–67.
158. Wagner G, Levin RJ. Human vaginal fluid, pH, urea, potassium and potential difference during sexual excitement. In: Gemme R, Wheeler CC, eds. Profess in sexology: Selected proceedings of the 1976 international congress of sexology. New York: Plenum Press; 1976:335–44.
159. Lavoisier P, Aloui R, Schmidt MH, Watrelot A. Clitoral blood flow increases following vaginal pressure stimulation. Arch Sex Behav 1995;24:37–45.
160. Kukkonen TM, Paterson L, Binik YM, Amsel R, Bouvier F, Khalifй S. Convergent and discriminant validity of clitoral color Doppler ultrasonography as a measure of female sexual arousal. J Sex Marital Ther 2006;32:281–7.
161. Maravilla KR. Blood flow: Magnetic resonance imaging and brain imaging for evaluating sexual arousal in women. In: Goldstein I, Meston CM, Davis SR, Traish AM, eds. Women's sexual function and dysfunction: Study, diagnosis and treatment. London: Taylor & Francis; 2006:368–82.
162. Kukkonen TM, Binik YM, Amsel R, Carrier S. Thermography as a physiological measure of sexual arousal in both men and women. J Sex Med 2007;4:93–105.

163. Suh DD, Yang CC, Cao Y, Heiman JR, Garland PA, Maravilla KR. MRI of female genital and pelvic organs during sexual arousal. J Psychosom Obstet Gynaecol 2004;25:153–62.

164. Maravilla KR, Cao Y, Heiman JR, Yang C, Garland PA, Peterson BT, Carter WO. Noncontrast dynamic magnetic resonance imaging for quantitative assessment of female sexual arousal. J Urol 2005;173:162–6.

165. Maravilla KR, Heiman JR, Garland PA, Yunyu C, CarterWO,Peterson BT,Weisskoff RM. Dynamic MR imaging of the sexual arousal response in women. J Sex Marital Ther 2003;29:71–6.

166. Deliganis AV, Maravilla KR, Heiman JR, Carter WO, Garland PA, Peterson BT, Hackbert L, Cao Y, Weisskoff RM. Female genitalia: Dynamic MR imaging with use of MS-325—initial experiences evaluating female sexual response. Radiology 2002;225:791–9.

167. Buisson O, Foldes P, Paniel B. Sonography of the clitoris. J Sex Med 2008;5:413–7.

168. Foldes P, Buisson O. The clitoral complex: A dynamic sonographic study. J Sex Med 2009;6: 1223–31.

169. O'Connell HE, Hutson JM, Anderson CR, Plenter RJ. Anatomical relationship between urethra and clitoris. J Urol 1998;159:1892–7.

170. Foldes P, Buisson O. Clitoris et point G: Liaison fatale. Gynecol Obstet Fertil 2007;35:3–5.

171. Basson R. Sexual desire and arousal disorders in women. N Engl J Med 2006;354:1497–506.

172. Davis SR, Davison SL, Donath S, Bell RJ. Circulating androgen levels and self-reported sexual function in women. JAMA 2005;294:91–6.

173. Dennerstein L, Lehert P, Burger H. The relative effects of hormones and relationship factors on sexual function of women through the natural menopausal transition. Fertil Steril 2005;84:174–80.

174. Grafenberg E. The role of urethra in female orgasm. International Journal of Sexology. 1950;3:145–148.

175. Puppo V. The G-spot does not exist. Response by V. Puppo to the article "O. Buisson: the Gspot and lack of female sexual medicine. Gynécologie Obstétrique & Fertilité 2010;38:781-84"

176. д-р Е. Дюпуи. «Проституция в древности». Издательство Кишинев "Logos" 1991. ISBN 5-85886-017-6.

177. «Поэзия Вагантов» под редакцией М.Л. Гаспарова, Издательство «Наука», Москва, 1975.

178. Publius Ovidius Naso. Ars Amatoria. // Публий Овидий Назон. «Наука любви. Лекарство от любви», составитель А. Марков. ООО Издательство «Эксмо», Москва, 2006. ISBN 5-699-15943-6.

179. Эрика Джонг. «Страх полета», Издательство «Эксмо»1994 г., ISBN: 5-85585-140-0

180. Frans de Waal, "Bonobo Sex and Society", *Scientific American* (March 1995) 82–86.

181. Bailey NW, Zuk M (August 2009). "Same-sex sexual behavior and evolution". *Trends Ecol. Evol. (Amst.)* **24** (8): 439–46. doi:10.1016/j.tree.2009.03.014. PMID 19539396.

182. Bruce Bagemihl, *Biological Exuberance: Animal Homosexuality and Natural Diversity* (St. Martin's Press, 1999). ISBN 0-312-19239-8

183. Corty, E., & Guardiani, J. (2008) Canadian and American Sex Therapists' Perceptions of Normal and Abnormal Ejaculatory Latencies: How Long Should Intercourse Last?. The Journal of Sexual Medicine, 5(5), 1251–1256.

184. "Sex therapists: Best sex is 7 to 13 min". Upi.com. March. 5, 2008.

185. Wayne Weiten, Dana S. Dunn, Elizabeth Yost Hammer (2011). Psychology Applied to Modern Life: Adjustment in the 21st Century. Cengage Learning. pp. 384–386. ISBN 1-111-18663-4, 9781111186630. Retrieved January 5, 2012.

186. "I Want a Better Orgasm!" WebMD. Retrieved August 18, 2011.

Brief vocabulary of terminology and abbreviations

A

Amenorrhea - absence of menstruation for more than 6 months
Androgens - male sex hormones (Testosterone)
Anesthesia - lack of sensitivity, pain relief
Anorgasmia - lack of orgasm during sex
Autoeroticism - self oriented sexual desire

B

Bisexuality - the presence of sexual attraction to persons of both sexes

C

Coital - occurs during sexual intercourse
Coitus - sexual intercourse
Coitofobiya - fear of sexual acts

D

D/S: Dominance and submission.
Defloration - technical virginity loss due to hymen rupture
Deviation - sexual deviations from generally accepted in the society forms of sexual behavior
Disgamiya - disharmony in sexual relations
Dyspareunia - painful intercourse

E

Ejaculation, - sperm emission
Extracoitally - out of sexual intercourse
Endocrine - caused by the action of the endocrine glands
Endogenous - caused by internal factors
Endometrium – mucosa in uterine cavity
Endorphins - morphine-like substances, "happiness hormone"
Erection – induration and straightening of penis
Erogenous - increasing sensitivity to sexual stimuli
Erogenous zones – sensitive areas of the body
Estrogen - the female sex hormones
Etiology - the cause of the disease
Euphoria - unmotivated high spirits, excitation
Excretion – discharge of products of metabolism (urine, sweat, etc.)

Exhibitionism - sexual perversion: the public exposure of the genitals for the purpose of sexual gratification

Exogenous - caused by external causes

F

Fetishism - the emergence of sexual arousal only in the presence of fetish, for example, women's braids, shoes, etc.

Friction - the friction of the penis against the walls of the vagina

Frigidity - 1) sexual coldness, and 2) the inability to experience orgasm.

Frustration - sexy painful condition caused by an acute sense of dissatisfaction with the fact that sexual arousal is not completed orgasm

G

Gene - structural and functional unit of heredity

Genesis - the origin, development

Genetic – hereditary transmissions

Genitalgia - pain in the genital area

H

Harmony - sexual pleasure to both partners

Heterosexuality - sexual attraction to persons of the opposite sex

Homosexuality - sexual attraction to persons of the same sex

Hormones - biologically active substances secreted by the endocrine glands

Hypersexuality - increased libido

Hypogonadism - reduced secretion of sexual hormones, leading to underdevelopment of sex organs and secondary sex characteristics

Hypothalamus - a department of an intermediate brain, the pituitary gland regulates the activity, metabolism, cardiovascular, sexual, digestive and other systems

Hypothyroidism - low function of thyroid

Hypo – decreased, low

Hyper – increased, higher

I

Iatrogenic - disorders caused by careless remarks doctor or his behavior

Incest - sex with blood relatives

Intsestofiliya - sexual perversion: sexual attraction to blood relatives

Inversion of sexual desire - desire to focus eponymous sex

252

Involution - an age-regression of

L
Lability - instability
Lesbian - female homosexuality
Libido - sex appetence
Limbic system - the system set of brain regions, located mainly on the inner surface of the cerebral hemispheres, involve emotions, memory, etc.

M
Masculinization - see virilization
Masochism - sexual satisfaction derived from sexual stimulation or humiliation
Masturbation - artificial sexual self-stimulation, sexual intercourse takes place outside
Menopause - the period of a woman's life when menstruation finally stopped

N
Narcissism - sexual perversion: to own self desire
Neurasthenia - neurosis, characterized by increased irritability and fatigue
Neurohumoral regulation - implemented through the nervous system and body fluids (blood, cerebrospinal fluid, etc.) by hormonal
Neuroleptics - a group of drugs that have an inhibitory effect on the nervous system
Neurosis - reversible disruption of nervous activity caused by traumatic effects and prolonged emotional stress
Nymphomania - increased sexual desire in women

O
Orgasm - higher voluptuous sexual feeling, ending sense of sexual satisfaction
Ovary - female gonad
Ovulation - release of an oocyte from the ovary

P
Penetration – introduction, intromission of penis into vagina (or anus, etc)
Per anum - through the anus

Per rectum - through the rectum

Perversion - a violation of sexual desire or conditions of its implementation

Petting - evoking orgasm foreplay without intercourse

Phobia - an obsessive fear

Pituitary gland - endocrine gland of the brain, located in the cranial cavity, produces a number of peptide hormones, regulates the activity of other endocrine glands

Plateau phase - the phase of sexual intercourse, with higher rates of sexual arousal

Pollution, Wet-dream - ejaculation occurs in men spontaneously during sleep

Potency - sexual power

Psychasthenia - a form of neurosis in patients with anxiety and doubtful character.

Psychogenic - mental effects caused by

Psychopathy - mental abnormality caused neuroses

Psychotherapy - treatment of mental techniques (persuasion, suggestion and hypnosis)

Puberty - the period of the signs of sexual maturity

R

Refractory period - the period nonexcitability

Reproduction - childbirth

Retardation - low development

S

S/M: Sado/Maso, Sadism and Masochism

Sadism - sexual satisfaction provided partner abuse

Satyriasis - increased sex drive in men

Secret - a product of the endocrine glands

Senestopatiya - unusual painful bodily feelings in psychiatric patients

Sensitive - touchy, vulnerable

Sensitization - increased sensitivity

Sexual Autoidentification - definition of belonging to the male or female

Sinton - empathy

Sperm - the male sex cell

SSC: *See* safe, sane, and consensual.

Suggestion - hypnosis

Symptomatic agents - addresses the diverse manifestations of the disease

Syndrome - a group of related symptoms.

Synergistic - friendly

T

Tactile - caused by touching

Testosterone - the male sex hormone

Transsexuals - people feel that they belong to the opposite sex, and thus seeking to change their gender surgically

Transvestism, transvestite - wearing and the desire to appear as the opposite sex

V

Vaginism - a reflex contraction of the muscles of vestibule and pelvic floor, hampering sexual intercourse or gynecological examinations

Virgogamiya - virgin marriage: a marriage in which the wife of a long time (months or years) can not start having sex

Virilization - the appearance of a woman's masculine traits caused by the action of androgens

Voyeurism - the kind of sexual perversion: desire to contemplate sexual intercourse or genital nudity

Vulva - the external female genital organs

Z

Zoophylia - sexual attraction to animals

Note page

Note page

ISBN - 978-1-300-74443-6

www.ingramcontent.com/pod-product-compliance
Lightning Source LLC
Chambersburg PA
CBHW061504180526
45171CB00001B/29